# EINSTEIN'S FRIDGE

HOW THE DIFFERENCE
BETWEEN HOT AND COLD
EXPLAINS THE UNIVERSE

## PAUL SEN

SCRIBNER

New York   London   Toronto   Sydney   New Delhi

Scribner
An Imprint of Simon & Schuster, Inc.
1230 Avenue of the Americas
New York, NY 10020

First Scribner hardcover edition March 2021

SCRIBNER and design are registered trademarks of The Gale Group, Inc.,
used under license by Simon & Schuster, Inc., the publisher of this work.

For information about special discounts for bulk purchases,
please contact Simon & Schuster Special Sales at 1-866-506-1949
or business@simonandschuster.com.

The Simon & Schuster Speakers Bureau can bring authors to your live event.
For more information or to book an event, contact the Simon & Schuster Speakers Bureau
at 1-866-248-3049 or visit our website at www.simonspeakers.com.

Manufactured in the United States of America

1   3   5   7   9   10   8   6   4   2

Library of Congress Cataloging-in-Publication Data has been applied for.

ISBN 978-1-5011-8130-6
ISBN 978-1-5011-8132-0 (ebook)

Image Credits
Original illustrations by Khokan Giri; page 91: Maxwell's apparatus to measure gas viscosity, with
kind permission and courtesy of the Cavendish Laboratory, University of Cambridge; page 151:
Brownian motion diagram from Jean Perrin's paper, *Journal de Physique, Theorique et Appliquée*
Volume 9, Numéro 1, 1910; page 163: one of Einstein and Szilard's refrigerator patents, Deutsches
Patent- und Markenamt; page 209: an example of a "dappled" pattern, as shown in Turing's
paper: *Philosophical Transactions of the Royal Society*, Biological Sciences; page 215:
fruit fly larvae, Shutterstock.

To Joseph and Nathan

# Contents

# Contents

# Prologue

*Thermodynamics* is a dreadful name for what is arguably the most useful and universal scientific theory ever conceived.

The word suggests a narrow discipline concerned only with the behavior of heat. Here indeed lie the subject's origins. But it's grown far beyond that and is now more broadly a means of making sense of our universe.

At its heart are three concepts—energy, entropy, and temperature. Without an understanding of these and the laws they obey, all science—physics, chemistry, and biology—would be incoherent. The laws of thermodynamics govern everything from the behavior of atoms to that of living cells, from the engines that power our world to the black hole at the center of our galaxy. Thermodynamics explains why we must eat and breathe, how the lights come on, and how the universe will end.

Thermodynamics is the field of knowledge on which the modern world is based. In the years since its discovery, we have seen the greatest improvement in the human condition in the history of our species. We live longer, healthier lives than ever before. Most children born today are likely to reach adulthood. Though much remains wrong with our time, few of us would swap places with our ancestors. Thermodynamics alone didn't cause this, but it was essential for it to happen. From sewage pumps to jet engines, from a reliable electricity supply to the biochemistry of lifesaving drugs, all the technology that we take for granted needs an understanding of energy, temperature, and entropy.

Yet despite its importance, thermodynamics is the Cinderella of the sciences. The subject is introduced piecemeal in secondary school physics, and the concept of entropy, so vital to our understanding of the universe, is barely mentioned.

I first encountered the study of thermodynamics in the second year toward my undergraduate engineering degree at Cambridge University, where it was presented as relevant only to car engines, steam turbines, and refrigerators. If instead I had been told that it provided a unified and coherent way of understanding all science, I might have paid more attention. Most adults are similarly introduced to the topic; even ones who consider themselves educated are ignorant of humanity's greatest intellectual achievement in the sciences. We count calories, pay energy bills, worry about the temperature of the planet, without appreciating the principles underpinning those actions.

The Cinderella status of thermodynamics is reflected in the way Einstein's science is remembered. All acknowledge his immense and revolutionary contributions, yet few realize the extent to which his work derived from thermodynamics or that he made seminal contributions to the subject. In his so-called miracle year of 1905, he published four papers that transformed physics, including the one featuring the equation $E = mc^2$. This work did not emerge from nowhere. For in the previous three years, Einstein had published three papers on thermodynamics, and the first two of the miracle-year papers—one on the atomic structure of matter and the other on the quantum nature of light—were continuations of that work. The third miracle-year paper, on special relativity, took an approach to physics inspired by thermodynamics, and the fourth, in which he derived $E = mc^2$, united the Newtonian concept of mass with the thermodynamic concept of energy.

Of thermodynamics Einstein said, "It is the only physical theory of universal content, which I am convinced . . . will never be overthrown."

Nor was Einstein's interest in thermodynamics limited to its role in fundamental and theoretical physics. He cared about its practical applications, too. In the late 1920s, he worked on designing cheaper and safer refrigerators than those available at the time. This little-known episode was not a quirky sideline, for he worked for several years on the project and successfully raised funding for it from the engineering companies AEG and Electrolux. The direct motivation for Einstein's interest in refrigerator design was that, in 1926, he read an article in a Berlin newspaper about a family—which included several children—who died because their malfunctioning refrigerator had leaked lethal fumes. Einstein's response was to initiate a project to design safer refrigerators.

Thermodynamics isn't just great science; it's great history, too.

• • •

In early 2012, while producing a television documentary, I came across *Reflections on the Motive Power of Fire*, a slim book self-published in Paris in 1824 by a reclusive young Frenchman called Sadi Carnot.

Carnot had died of cholera at thirty-six, believing that his work would be forgotten. Yet within two decades of his death, he was considered the founding father of the science of thermodynamics. Later in the nineteenth century, the great physicist Lord Kelvin said of Carnot's text, "that little essay was indeed an epoch-making gift to science."

I also became captivated. Carnot's work was unlike any other work of fundamental physics, combining algebraic calculus and physical insight with Carnot's thoughts on what would constitute a happier, fairer society. Caring deeply for humanity, Carnot believed science was the key to progress.

Carnot's science was also a response to the seismic social changes in early nineteenth-century Europe. In that sense, *Reflections* was as much the product of two revolutions—the French and the Industrial—as it was of Carnot's brilliant mind. As I then started to read more about the scientists who picked up the baton from him, I saw how all their work was influenced by events in the world around them. The story of thermodynamics is not only one about how humans acquire scientific knowledge, it is also about how that knowledge is shaped by and, in turn, shapes society.

This book is an argument that the history of science is the history that matters. The men and women who push back the frontiers of knowledge are more important than generals and monarchs. In the following pages, I shall therefore celebrate the heroes and heroines of science and show their quest to discover the truth about the universe as the ultimate creative endeavor. Sadi Carnot, William Thomson (Lord Kelvin), James Joule, Hermann von Helmholtz, Rudolf Clausius, James Clerk Maxwell, Ludwig Boltzmann, Albert Einstein, Emmy Noether, Claude Shannon, Alan Turing, Jacob Bekenstein, and Stephen Hawking are among the smartest humans who ever lived. To tell their story is a way for all of us to comprehend and appreciate one of the greatest achievements of the human intellect.

Ludwig Boltzmann, one of the heroes of this story, put it this way:

"It must be splendid to command millions of people in great national ventures, to lead a hundred thousand to victory in battle. But it seems to me greater still to discover fundamental truths in a very modest room with very modest means—truths that will still be foundations of human knowledge when the memory of these battles is painstakingly preserved only in the archives of the historian."

# CHAPTER ONE

# A Tour of Britain

The number of steam engines has multiplied prodigiously.
—French economist and businessman
Jean-Baptiste Say on visiting Britain

On September 19, 1814, Jean-Baptiste Say, a forty-seven-year-old French businessman and economist, embarked on a ten-week spying mission to Britain. Napoléon had been exiled to the Mediterranean island of Elba three months earlier, and the trade blockade between France and her northern neighbor had ended. The new government in Paris sensed an opportunity to investigate the reasons underpinning Britain's recent economic surge, and in Jean-Baptiste Say, they found the ideal man. Say had lived for two years in Britain as a teenager, working in the offices of various British trading companies and learning fluent English. Later, he'd run a textile factory in northern France and become a published economist, thus acquiring both a practical and theoretical appreciation of commerce.

As spying missions go, Say's was neither dangerous nor clandestine. He made no secret of his reasons for being in Britain. A gregarious Anglophile, he crisscrossed the country, obtaining access to mines, factories, and ports and, in his leisure time, to theaters and country houses. And since his last visit twenty-six years earlier, Say witnessed a nation transformed. He began his tour in Fulham, a village to the west of London where he'd spent time in his youth. He found it unrecognizable. There were new houses all around, and a meadow he'd enjoyed strolling through years before had become a shop-filled street.

For Say, Fulham's metamorphosis was representative of what had hap-

1

pened over the eighteenth century to the country as a whole. Britain's population had soared, growing from 6 million to 9 million, and her people
had become the best fed, clothed, and paid in Europe. Trade had bur-
geoned, too—Say noted that the number of ships in the port of London
had tripled to three thousand. In other parts of the country, he admired
new canals and city streets illuminated by gaslighting. He took in a foundry
for machine parts in Birmingham, a seven-story textile spinning factory
in Manchester, coal mines near York and Newcastle, and a steam-powered
mill for weaving cotton fabrics in Glasgow. Its owner, a certain Finlay, was
so proud of this machinery and indeed so unperturbed at the thought of
potential French competition that he showed Say how it worked himself.

Powering this economic miracle was Britain's cotton-manufacturing
industry, whose export value had shot up twenty-five-fold in the time
between Say's first visit in the 1780s and his second in the 1810s. Many in
France, including those who had had Napoléon's ear, believed that the best
way to emulate this was by acquiring an empire—Britain, after all, had
access to cheap raw cotton from her colonies. Say disagreed. He considered colonialism to be unprofitable in the long run and instead regarded
technological innovation as the key to Britain's success. Above all else, one
piece of technology caught Say's eye and his imagination:

"Everywhere, the number of steam engines has multiplied prodigiously.
Thirty years ago, there were only two or three of them in London; now
there are thousands. . . . Industrial activity can no longer be profitably sus-
tained without the powerful aid they give."

Above all, steam power had revolutionized Britain's mining industry.
Mines, like water wells, are shafts dug into the ground and are prone to
flooding. The preindustrial horse-driven pumps had struggled to lift water
out of any mines that were more than a few yards deep. Moreover, it takes
around two acres to feed a horse for a year, meaning there wasn't enough
grazing land in Britain to feed the number of horses widespread mining
would require. But by 1820, steam technology had advanced to the point
where engines could easily pump water out of shafts that were over three
hundred yards deep. This lowered the cost of mining coal, which, because
coal is a crucial ingredient in the manufacture of iron, made iron more
abundant, too. Between 1750 and 1805, production of the metal soared
ninefold from 28,000 to 250,000 tons a year.

•  •  •

Steam power in early nineteenth-century Britain was ubiquitous but not as innovative as Say thought. The technology had proliferated not because Britons were especially inventive, but because their country was so replete with coal that even poorly designed and wasteful engines were profitable. Take, for example, the one installed at the Caprington Colliery in southwest Scotland in 1811, which operated on a principle pioneered a century earlier by an English inventor called Thomas Newcomen. Devices such as this weren't what we, in the twenty-first century, regard as steam engines, in which the pressure exerted by hot steam pushes a piston. Instead, they are best understood as steam-enabled vacuum engines. The relationship between the heat created in their furnaces and the mechanical work they perform is convoluted and inefficient.

A Newcomen engine

"Newcomen engines" work as follows: Heat from burning coal creates steam. This flows via an inlet valve into a large cylinder in which a piston can move up and down. Initially the piston rests at the top of the cylinder. Once this is full of steam, the inlet valve closes. Cold water is sprayed into the cylinder, cooling the steam inside, causing it to condense into water. Because water occupies much less space than steam, this creates a partial vacuum below the piston. Atmospheric air will always try to fill a void, and the only way it can do so in this arrangement is by pushing the piston down. This is the source of the engine's power. The steam is a means to create a vacuum and the downward pressure of the atmosphere does the work.

To observe this effect, pour a small amount of water into an empty soft-drink can and warm it until it's filled with steam. Take some safety precautions and pick up the can with tongs—it will be hot—and quickly turn it upside down as you submerge it in a bowl of ice-cold water. The steam condenses into water, thus creating a partial vacuum inside the can. Pressure from the earth's atmosphere will then crush the can.

In the steam engine I've been describing, this process—filling the cylinder with steam and condensing it to water so a partial vacuum is created—repeats over and over. Thus, the piston goes up and down, powering a pump.

Newcomen engines consumed prodigious amounts of coal. They burned a bushel—84 pounds—of coal to raise between 5 to 10 million pounds of water by one foot. This quantity, the amount of water that can be raised by one foot for every bushel burned, was called the engine's *duty*. By modern standards, these engines were very inefficient, wasting around 99.5 percent of the heat energy released as the coal burned.

That such wasteful engines continued to be used for over a century was due to cheap coal. At the time of Say's visit, Britain's mines produced 16 million tons every year, and in the new industrial towns of Leeds and Birmingham, coal often sold at less than ten shillings per ton. At these prices, poor engine design mattered little.

Then in 1769, the Scottish engineer James Watt had patented a modification to the Newcomen engine, which roughly quadrupled its duty. But the arrival of Watt's designs, paradoxically, put a brake on British innovation for thirty years as he and his business partner Matthew Boulton used the patent system to prevent other engineers from bringing further improvements to the market. Then, as now, commercial success was not necessarily aligned with innovation.

In addition, the people of England had a love-hate relationship with science. On the one hand, over the eighteenth century, the country's growing middle class had developed a great interest in natural philosophy, as science was termed. Encyclopedias were bestsellers. Crowds flocked to public lectures that covered topics from the behavior of magnets to recent astronomical discoveries. Clubs sprang up as informal gatherings for scientific discussion. The most famous came to be known as the Lunar Society, which counted Watt and Boulton as members. But on the other hand, some sections of the public also grew wary of science because many of its practitioners, such as Joseph Priestley, the discoverer of oxygen, publicly supported the radical politics of the French Revolution. He paid dearly for his views. In 1791, an angry mob burned down his house and laboratory.

Moreover, England's two universities, Oxford and Cambridge, offered no courses in subjects that resemble modern-day physics and engineering. Cambridge, being Isaac Newton's alma mater, did rigorously train students in the mathematical principles that great scientist had discovered. But basking in Newton's legacy, professors there saw no need to extend his work and were suspicious of novel mathematical techniques being developed abroad. In 1806, when one progressive scholar, Robert Woodhouse, urged the adoption of a European style of mathematics, he was condemned as unpatriotic in the conservative *Anti-Jacobin Review*. The real-world applications of mathematics were also not a priority. Yes, Newton's laws did describe aspects of the universe we inhabit such as the orbits of planets. But Cambridge professors felt the purpose of teaching the laws was to provide mental training to students drawn from the landed gentry who would go on to serve church, state, and empire. Cambridge students railed against this, but it would be decades before attitudes changed.

France, however, was very different.

Jean-Baptiste Say published his observations on Britain's economic and industrial transformation in a book entitled *De l'Angleterre et des Anglais*, in 1816. His report, and those of others, convinced French engineers, businessmen, and politicians that the way to catch up with Britain economically was to exploit steam power. But they faced a problem: coal was scarce south of the Channel. French mines produced a million tons annually, and as most of these were in the remote Languedoc region, the price never dropped below twenty-eight shillings per ton, three times higher than in

England's industrial heartland. This meant that from the earliest stages of their country's industrialization, French engineers cared about engine efficiency—how to maximize the useful work that can be extracted from burning a given amount of coal—in a way most of their British counterparts did not.

French scientific and mathematical education was also very different from that in Britain, as is exemplified by the institution where Say became professor of industrial economy three years after returning to his homeland. The National Conservatory of Arts and Crafts, as it was named, was a far cry from an elite institution such as Cambridge. Located in Paris, the Conservatory was created as part of the French revolutionary government's commitment to public education, and it embodied that regime's conviction that science and mathematics were weapons in a war against superstition and arbitrary aristocratic privilege. They provided rational laws to help found a rational society. Subsequently, Napoléon continued to support these subjects, seeing them as important to France's military ambitions. Working in this context, French scientists, therefore, saw Newton's work as a foundation on which to build. They widened its reach and made it far simpler to use. At places such as the Conservatory, it was natural to think that mathematical analysis could be applied to steam engines and, in particular, to their efficiency.

And here a young student laid the foundations of the science of thermodynamics.

# CHAPTER TWO

# The Motive Power of Fire

It is necessary that there should also be cold; without it, the heat would be useless.

—Sadi Carnot

The young man is extremely gentle, he behaves well and is a little shy. . . . His confidence must not be undermined.

—A friend's description of Sadi Carnot

Still in his twenties, of medium build and possessing a "delicate constitution," Sadi Carnot was reserved and introspective and lived a solitary life. Fellow students at the Conservatory of Arts and Crafts in Paris in the early 1820s paid him little heed. A surviving portrait pictures him as cultured, thoughtful, and yet somehow fragile in appearance.

Sadi Carnot was born on June 1, 1796, in a room in the Palace of the Petit Luxembourg in Paris. His father, Lazare, was a gifted mathematician and engineer, who as a young man had published a paper suggesting ways of improving the Montgolfier brothers' famous hot-air balloon of 1783. Lazare's other scientific essays included investigations into the principles underpinning machines such as water mills. Lazare was also an admirer of a thirteenth-century Persian poet, Saadi of Shiraz, hence the unusual first name he had given his son.

In 1789, when the French Revolution began, Lazare turned to politics, and two years later, he won election as a deputy to the country's quasi-democratic Legislative Assembly. He then rose to prominence thanks to his highly effective reorganization of the French Revolutionary Army.

Lazare enjoyed a fair share of luck, too, surviving the Terror, unlike many other leading French revolutionaries. So when Sadi was born in 1796, his father was one of the five-member Directory, which ruled France, meaning the child was brought up at the epicenter of the greatest political and intellectual upheaval in eighteenth-century Europe.

Lazare Carnot himself educated Sadi as a boy, but when his son's aptitude for science became apparent, he sent him to France's preeminent center for scientific higher education, the Polytechnic School in Paris. Like the Conservatory of Arts and Crafts, which Sadi Carnot would attend later in life, the Polytechnic School had been created in 1794 as part of the French revolutionary government's commitment to public education. (Lazare Carnot was one of the founders.) The school's selectors traveled throughout France, aiming to find the country's most talented candidates, irrespective of their families' wealth. This worked to some extent, but overall, the school's intake was mostly from the upper classes. Its entrance examination was tough, and the best way of passing it was to receive training from an elite Parisian lycée or to be privately tutored, as Carnot was. He enrolled in November 1812, the third-youngest applicant that year at the age of sixteen; Carnot ranked 24 out of a field of 184.

At the Polytechnic, Carnot received two years of exemplary training in the latest discoveries in mathematics and physics, graduating in October 1814. He was destined for a career in the engineering corps of the French military when history intervened. On June 18, 1815, British, Prussian, and other allied European forces defeated Napoléon at Waterloo and banished him to the remote mid-Atlantic island of St. Helena. Over a million foreign troops, the so-called Army of the Seventh Coalition, then occupied France and enthroned a new king, Louis XVIII, brother of Louis XVI, who had been beheaded during the revolution. These events proved calamitous for the Carnot family, not least because Napoléon had appointed Lazare Carnot minister for the interior shortly before his defeat. Such closeness to Napoléon meant that the post-Waterloo French regime distrusted Lazare, and as a result it exiled him to the town of Magdeburg in Germany. Remaining in Paris, Sadi Carnot found himself treated as a pariah. During Napoléon's rule, high-ranking French soldiers would seek out Sadi and flatter him because he bore the Carnot name; now he found himself shunned and dispatched by his new military superiors to remote parts of France. It must have been a huge relief to Carnot that in 1819 he secured a posting as a lieutenant back in Paris, was put on half pay, and, apart from occasional military training exercises, was left to his own devices.

Carnot used his free time to cultivate his interest in science and technology. He visited factories in the newly developing industrial areas of Paris and enriched his earlier scientific education by attending lectures at the Conservatory of Arts and Crafts, where Jean-Baptiste Say was teaching. Located in the east of Paris, it had been housed by the revolutionary government in a repurposed monastery and like the Polytechnic School, its mission was to further public education. The restored Bourbon government continued to fund the Conservatory but, because of its association with previous regimes, suspected many of its lecturers and its students of secretly plotting rebellion and infiltrated the institution with spies.

Nonetheless there was an exciting spirit of inquiry at the Conservatory, and here Carnot encountered its professor of chemistry, Nicolas Clément, who taught him all that was known about temperature and heat.

Of the two, temperature is the easier concept. For an intuitive grasp in line with early nineteenth-century views, think of temperature as a measure of how hot something feels. Imagine, for example, a large pot and a small saucepan. Both have been filled with water from the same tap. Placing your finger in either produces a similar sensation. Placing a thermometer in either will show the same reading.

Heat in much trickier to understand. Place the same two vessels on a stove, and the temperature of the water they contain goes up as "heat" is released from the burning gas. But to get the same hike in temperature, you must place the larger vessel on the stove for a much longer time than the smaller one. These observations imply that the effect of heat on a substance is to raise its temperature by an amount that depends on the quantity of the substance. But what is heat? What is emanating from the burning gas that makes things hotter?

In Clément and Carnot's day, most scientists believed that heat was an invisible substance called caloric, made up of tiny, weightless particles released from within burning substances. These caloric particles, it was supposed, repelled one another, and this was why heat tended to spread from hot to cold, equalizing temperature differences. As the particles of caloric pushed away from one another, they seeped through the tiny pores that were believed to exist in all materials, diffusing through them and thus making them hotter. The larger the volume of the substance, the more caloric was needed to cause a given temperature increase. Caloric didn't only make things hotter—it could cause them to melt or boil. Many scientists regarded caloric as a gaseous element like oxygen, which could flow from one place to another. And just as elements such as oxygen could not be created or destroyed, neither could caloric.

By the early 1800s, however, many scientists grew aware of weaknesses in caloric theory. One such was an American émigré scientist based in Munich named Benjamin Thompson, working as aide-de-camp to the ruler of Bavaria. His duties included overseeing the national arsenal, and he observed that when cannon barrels were hollowed out by a tool resembling a giant drill bit, the friction generated an enormous amount of heat. To investigate further, Thompson immersed a cannon barrel in water while it was being drilled. After two and half hours, so much heat was generated that the water began to boil.

In a paper submitted to Britain's leading scientific body, the Royal Society, Thompson argued that though caloric theory could explain why heat was released from burning, it couldn't explain why it was released by friction. In the former process, it was plausible that trapped caloric particles escaped as fuel burned. Once the fuel was used up, caloric was no longer released. Friction, on the other hand, appeared to be a limitless source of heat. As long as mechanical effort was spent rubbing two objects together, heat would emerge. In other words, friction seemed to *create* heat, not release it. This went against the assertion in caloric theory that heat could neither be created nor destroyed. (Thompson, archcritic of caloric theory, married Marie-Anne Lavoisier, widow of one of the theory's founders, the famous French chemist Antoine Lavoisier, who had been executed during the Terror. Thompson and Mme Lavoisier's marriage was short.)

In addition to the strengths and weaknesses of caloric theory, Carnot learned Clément's own contribution to the study of heat, namely that he had devised an objective way of quantifying it. Prior to Clément, despite people's building steam engines for over a century, there was no universally agreed unit for measuring amounts of heat. Cornish mining engineers had come up with the concept of an engine's "duty"—the weight of water in pounds raised by one foot when a bushel, or eighty-four pounds, of coal was burned in its furnace. But they hadn't thought to quantify the heat given off by the coal as it burned. People also knew, for instance, that it took more heat to boil a liter of water than it took to boil a liter of alcohol, but there was no agreed way to compare the different amounts of heat numerically. Clément came up with a method for doing so.

We know all this from an anonymous account of Clément's lectures that survives. In them are the historic words "Mr. Clément imagines a unit of heat that he names the 'calorie.' One calorie is the amount of heat needed to elevate by one degree centigrade one kilogram of water." That is still what a calorie means when used to measure the energy content of food. So for example, a hundred-gram packet of potato crisps that contains around five hundred calories, per Clément's definition, will release enough heat on burning to raise the temperature of five hundred kilograms of water by one degree Celsius. (A few decades later, scientists redefined the calorie to mean the amount of heat needed to raise the temperature of one gram of water, rather than one kilogram of water, by one degree Celsius, which means that one of Clément's calories is equivalent to one thousand calories now.)

Another influence on Carnot was his father Lazare's scientific papers, written in the decade before the revolution. In one entitled "An Essay on Machines in General," Lazare had mathematically analyzed the behavior of water mills.

Specifically, Lazare imagined an *ideal* mill in which all the "pushing power" of the water is turned into the rotary motion of the wheel and none is wasted. In such a mill, the water slows gradually as it turns the wheel, transferring all its speed of flow to the wheel's rotational movement. Lazare observed that real mills fell far short of this ideal, but he offered meager advice on how to remedy this. He focused, instead, on the physics underpinning waterpower with the aid of mathematics. Unsurprisingly mill builders paid little attention to Lazare's abstract form of reasoning, but his son would use this approach to great scientific effect.

In 1821, Carnot traveled to Magdeburg to visit his exiled father and younger brother for a few weeks. The timing was propitious. The city's first steam engine had been installed three years earlier by an expatriate English engineer—such men built a large proportion of the few engines in continental Europe at this time. It's not a stretch to speculate that Lazare and Sadi visited the engine and noted how the British led the world in steam technology. In any event, when Sadi Carnot returned to Paris, he set to work immediately on a seminal text. When he completed it in 1824 he called it *Reflections on the Motive Power of Fire and on Machines Fitted to Develop That Power*. By *motive power*, Carnot meant the amount of useful work, such as pumping water out of a mineshaft or powering a ship, that can be obtained from the heat that's created in the "fire" or furnace of a steam engine.

Carnot's text is nothing like a modern scientific paper. His desire that it should be "understood by persons occupied with other studies"—by which he meant nonscientists—shines through in its jargon-free, lucid exposition. Before explaining the science, Carnot tries to persuade the reader that the science matters. He stresses the benefits of the way steam engines use heat to perform tasks that hitherto had required animal-muscle power, wind, or flowing water, writing, "They seem destined to produce a great revolution in the civilized world." He even makes the case for the technology's utopian potential: "Steam navigation brings nearer together the most distant nations. It tends to unite the nations of the earth as inhabitants of one country." And as proof of what steam power is capable of, Carnot points across the English Channel: "To take away today from England her steam engines . . . would be to ruin all on which her prosperity depends . . . to annihilate that colossal power."

Carnot ends his introduction with this statement of intent: "Notwithstanding the work of all kinds done by steam engines . . . their theory is very little understood, and the attempts to improve them are still directed almost by chance."

For Carnot, deducing the theory underpinning steam engines was therefore no academic exercise. Doing so, he felt, would provide a way of improving their fuel efficiency, thus reducing costs for his country's industrialists and helping them catch up with their British counterparts. To Carnot, the crucial question was, How does one obtain as much motive power as possible from a steam engine?

Carnot then takes the idea of an engine's duty one step further. Instead of asking how much coal must be burned to raise a known weight a given distance, Carnot asks how much heat must flow out of a furnace to achieve this. Or put another way, if, say, one hundred calories of heat flow out of a furnace, what's the greatest possible height to which this can raise a weight of one kilogram? (For simplicity's sake, think of one unit of "motive power" as the amount that will lift a one kilogram weight by a height of one meter.)

To answer this question, Carnot considers a typical early nineteenth-

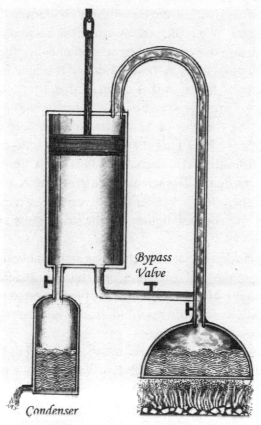

Impressionistic view of key aspects
of a Watt engine

century steam engine that works along the lines devised by James Watt. Two aspects greatly interested the Frenchman.

First, Watt had noticed that hot steam exerts a great deal of pressure, more even than the downward weight of the atmosphere. To exploit this, he contrived his engine design so expanding steam from a boiler pushed a piston. (In the diagram, the steam pushes the piston down.)

Second, Watt understood that for the engine to keep going, the piston must return to its starting position at the top of the cylinder. This requires the steam that has pushed it down to be cooled and condensed into water, so it no longer presses down on the piston. Then a portion of the motive power generated in the downward stroke is used to push the piston back up.

Watt enabled this with a device called a condenser that's kept cool by a spray of water. When the piston is near the bottom of the cylinder, the bypass valve and the valve leading to the condenser open. The steam above the piston flows through these valves and is turned into water in the condenser. It thus no longer pushes down on the piston.

In his treatise, Carnot ignores how the engine's components work and instead focuses on the flow of heat through the whole device. Adhering to caloric theory, he argues that a quantity of indestructible caloric fluid, released from the burning coal in a furnace, is "incorporated" into the steam, thus raising its temperature and pressure so it can push down on the piston. Then, in the condenser, the caloric is removed from the steam, causing it to cool and liquefy. As the steam pressure falls, the piston returns.

Carnot's conclusion is that an unchanging amount of caloric flows from the hot furnace to the cold condenser, and this flow generates the motive power of the engine. He equates this to the flow of water. Just as no water is lost as its downward flow drives a mill, so, too, no heat is lost as its "cool-ward" flow drives a steam engine.

Carnot was wrong to believe in caloric theory, but it did lead to his first insight. A body of water, no matter how vast, will not produce motive power unless it can flow downhill. So, too, even a prodigious quantity of heat will not create motive power if there's no temperature difference it can "flow down." A steam engine inside a vast hot furnace will not work despite the presence of abundant heat because there is no way to cool and liquefy the steam so the piston can be pushed back to the top of the cylinder. Carnot writes:

"The production of heat alone is not sufficient to give birth to the impelling power: it is necessary that there should also be cold; without it, the heat would be useless."

That sentence marks the first step in the history of thermodynamics.

Next, Carnot tackles a question that vexed engineers in his day: Is steam the best substance to use in machines that extract motive power from heat? After all, heating any gas, not just steam, causes it to expand, increasing the pressure it exerts. That means any gas can push a piston. So, can a machine based, say, on atmospheric air or alcohol vapor produce more motive power from a given amount of heat flow than a steam engine? Can such a machine use the heat flowing from burning one kilogram of coal raise a given weight to a greater height than a steam engine could?

In pursuit of an answer, Carnot ignores all the detailed engineering that goes into an actual machine. Instead, using a strategy learned from his father, he considers imaginary machines.

Carnot asks his reader to picture an *ideal* steam engine—one that, for a given flow of heat from a hot place to a cold one, can produce the *maximum* amount of motive power possible, i.e., it can lift a given weight to the greatest possible height. (For simplicity, I'll refer to the source of the heat as the *furnace* and the cooler place where the heat ends up as the *sink*.)

Next, Carnot proposes a hypothetical machine that does the same process *in reverse*—i.e., it uses up motive power to move heat from a cool place to a warmer one. In the modern world, we call such devices heat pumps or refrigerators. Again, Carnot is not interested in technical details. He reasons that if the flow of heat from a hot place to a cooler one can create motive power and raise a weight, then a machine can exist that does the opposite. In this machine, the motive power obtained from a falling weight will force heat to flow "uphill" from cool sink to hot furnace. There is a direct analogy with water mills and water pumps. The former uses the downward flow of water to produce motive power; the latter uses power to push water uphill.

Following Carnot's logic, now imagine an ideal engine, which on receiving one hundred calories of heat from a furnace, raises a fifty-kilogram weight by ten meters and then releases the heat into a sink.

Next, imagine a "reverse" ideal engine, in which, as a fifty-kilogram weight falls ten meters, one hundred calories of heat is taken up from the sink and put into the furnace.

With these two hypothetical machines defined, ask yourself, What happens if they are joined together? Now, the motive power produced by the ideal "forward" engine *drives* the ideal "reverse" engine.

Heat flow from furnace to sink through the ideal engine lifts a weight.

The weight is then connected to the ideal reverse engine and allowed to fall, "lifting" heat from sink to furnace.

This arrangement will run forever. One hundred calories of heat flow from the furnace into the forward engine and end up in the sink, lifting the fifty-kilogram weight along the way. The weight is connected to the

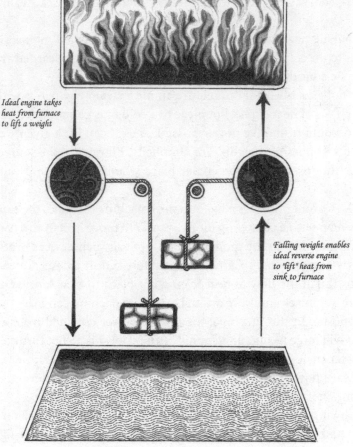

*Ideal engine takes heat from furnace to lift a weight*

*Falling weight enables ideal reverse engine to "lift" heat from sink to furnace*

An ideal engine drives an ideal reverse engine.

reverse engine and allowed to fall. This lifts one hundred calories of heat back up from the sink and replenishes the furnace with it. This fuels the forward engine, which lifts the weight back up and the whole loop repeats.

The point is that this furnace will never lose any heat and the weight will go up and down in perpetuity. But—and this is an important *but*— this system won't provide any usable motive power. The motive power produced by the forward engine is entirely consumed by the reverse engine. None is left over with which to do anything useful such as pump water.

Carnot's next step is a touch of genius. He conjures up another hypothetical engine that uses an alternative gas to steam, such as air or alco-

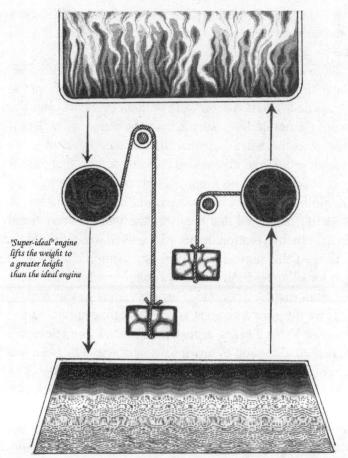

*"Super-ideal" engine lifts the weight to a greater height than the ideal engine*

A "super-ideal" engine drives an ideal reverse engine.

hol vapor; but the imagined gas he proposes using is *better* than steam. Thus, the engine using it is better than a steam engine, too. How much better, though? Suppose it lifts the fifty-kilogram weight by twelve meters rather than ten for the same flow of one hundred calories of heat between the same furnace and sink.

As before, Carnot analyzes the arrangement in which this "super-ideal" (nonsteam) engine drives the ideal reverse engine. One hundred calories flow into the super-ideal engine. But because it lifts the weight by twelve meters, this can drive the "reverse" engine *and,* say, a water pump. The reverse engine only needs the weight to fall by ten meters to send the hundred calories back to the furnace and replenish it. But with the super-ideal engine, the weight can still drop by another two meters at the end of each loop—twelve as opposed to ten.

This "leftover dropping distance," so to speak, could also be used to pump water. In fact, every loop will generate a surplus of "dropping power." Such a machine would do useful work without ever consuming any fuel.

However, Carnot declares, such a machine cannot exist. It is a perpetual motion machine, which scientists had long declared an impossibility. For centuries, people had dreamed of building devices that did something useful, but which needed no input of effort from animal muscle, flowing water, or wind. None had ever worked, and in 1775 the Royal Academy of Sciences in Paris stated that they would no longer consider proposals concerning perpetual motion. In his text analyzing water mills, Sadi's own father had used this same assumption of the impossibility of perpetual motion to set an upper limit on the amount of useful work that could be extracted from such devices. Sadi Carnot's genius was to realize that the same logic would work with great effect on steam engines. As he writes:

"Do we not know besides, a posteriori, that all the attempts made to produce perpetual motion by any means whatever have been fruitless?— that we have never succeeded in producing a motion veritably perpetual, that is, a motion that will continue forever without alteration in the bodies set to work to accomplish it?"

The impossibility of perpetual motion means, to Carnot, that it is not possible to build an engine that produces more motive power than an ideal steam engine. Or as he puts it:

"Such a creation is entirely contrary to ideas now accepted, to the laws of mechanics and of sound physics. It is inadmissible. We should then conclude that the *maximum of motive power resulting from the employment of steam is also the maximum of motive power realizable by any means whatever.*"

Carnot italicized the conclusion here himself. Far from saying that steam is the best material to use, he's instead arguing that *all ideal engines perform equally well, irrespective of the gas or any working material they use and how they are constructed.* The ideal steam engine may well look very different to an ideal air-based engine, but when working between the same furnace and sink, they will all lift a weight by the same amount. And that means that the working materials in the engine, such as the steam or the air, don't themselves provide any motive power—*that all comes from the heat flow*.

Carnot had considered steam engines in their idealized form and uncovered a truth about their real-world equivalent that no engineer had. Most still felt that the working material used by an engine did play some role in creating motive power.

For a sense of Carnot's reasoning, think of a water mill. For a given flow of water, the maximum power it can produce is limited by the height by which the water drops. No amount of cunning design can improve on this limit. The only way to up the power of the mill is to increase the height that the water drops. Analogously, for any heat engine, the power it can produce from a given flow of heat is limited by the temperature difference between its furnace and sink. The only way to up this is to increase this temperature difference. Conversely, reducing the temperature difference will reduce the power output.

Carnot also analyzed how to maximize the motive power produced by the flow of heat down any given temperature drop. In a typical engine, heat causes a gas like steam to expand and push a piston. In an ideal engine all the heat should go to expanding the gas and not be lost by, say, leakages. (For more details see Appendix I.)

This logic told Carnot that the real steam engines of his day had to be woefully wasteful. The hottest temperature the steam reached as it expanded and pushed a piston was, Carnot reckoned, a little over 160°C. The coldest it fell to as it condensed was around 40°C. That meant steam engines were extracting motive power from a temperature drop of around 120°C. But the temperature in the engine's furnace in which the coal was burning was over 1,000°C, and that meant a much-larger temperature drop—of 900°C or more—was being wasted.

The water mill is again a helpful analogy. Imagine a waterfall with a ten-meter drop. Now picture a waterwheel positioned only one meter below the top rather than at the bottom. Intuitively, one concludes that much of the power of the flowing water is being wasted. Steam engines waste heat flow in a similar way.

How could this be corrected? One way, Carnot argues, is to use atmospheric air as the substance that pushes the piston. Because air contains oxygen, fuel can burn and generate heat inside the cylinder and not in an external boiler as happens in a steam engine. "Considerable loss could thus be prevented" is how Carnot puts it. Air has another advantage—it has a lower "specific heat" than steam. That means, roughly, that the same amount of heat can raise the temperature of a quantity of air by a greater amount than an equivalent quantity of steam. In turn that implies that the same flow of heat can drive an air-based engine between greater temperature differences than a steam-based one. Thus, even more efficiency is achieved. Carnot writes, "The use of atmospheric air for the development of the motive power of heat . . . would doubtless offer a notable advantage over the vapor of water." This prediction was borne out in the late nineteenth century by the arrival of the internal combustion engine, a device that burns petrol or diesel to raise the air temperatures in its cylinders to well over 1,000°. Rudolf Diesel, who published his theories on how to build such an engine in 1893, was inspired by Carnot's ideas.

Carnot's treatise is a magnificent work of science, the product of a fertile imagination that worked in tandem with a mind that reasoned carefully on the basis of evidence. This legacy surrounds us. Internal combustion engines, jets, the giant turbines that generate electricity, and even the rockets that took humans to the moon, all are based on Carnot's discovery that the flow of heat from hot to cold is required to generate motive power. Less obvious, but just as important, is the legacy of Carnot's work to the quest to better understand our universe.

In the summer of 1824 Carnot published *Reflections on the Motive Power of Fire* at his own expense. He was twenty-eight years old. A wiser option might have been for him to submit his work to the widely read journal of his alma mater, the Polytechnic School. But perhaps the style of the *Reflections*, combining sociology, politics, and abstract reasoning, made it unsuitable? In any event, he self-published at a cost of 459.99 francs, which must have been a financial challenge given that he was living on half pay from the French military. Six hundred copies were printed and went on sale on June 12, 1824, at a price of three francs a copy. There is no account of how many were sold. Later that month, however, a summary of his book's ideas was read at a meeting of the Academy of Sciences in Paris, but there is no record of France's leading scientists having any recollection of that presentation and no evidence that Carnot was there to champion it.

Thereafter in the late 1820s, we catch glimpses of Sadi Carnot in the turmoil of French politics. In 1828, for instance, he resigned from the

French military, and following this there's no sign he had any form of employment, except for a letter indicating that he tried to go into business as an engineer.

In the years after the publication of *Reflections on the Motive Power of Fire*, there's a hint that Carnot might have lost faith in his work. Although few of his personal papers survive, one collection, a loose bundle of twenty-three pages, was found by his younger brother. Entitled "Notes on Mathematics, Physics, and Other Subjects," they reveal that Carnot was having doubts about a key assumption in the *Reflections*, namely that heat was an indestructible fluid known as caloric. In reflecting on when heat does something noticeable, such as pushing on a piston, he writes, "The quantity can no longer remain constant." With hindsight, we can read this concern as evidence of impeccable scientific instincts. But to Carnot, it would have cast doubt on his key insight, namely that without cold, heat is useless. Moreover, if heat isn't caloric fluid producing motive power as it flows from hot to cold, his entire hypothesis looks shaky. As Carnot put it in his notes, "It would be difficult to explain why, in the development of the motive power of heat, a cold body is necessary." This problem of how to reconcile Carnot's insight that heat must flow from hot to cold to produce motive power with this imaginary caloric fluid provides the next turn in our story.

Tragically, though, Carnot would play no further part. In 1832, for reasons that remain obscure, Carnot entered a mental asylum in Ivry, on the outskirts of Paris. While he was there, a cholera epidemic swept through France, and Carnot succumbed. Our last glimpse is of Sadi Carnot ravaged by a high fever and in mental anguish, dying with no knowledge of the immense importance of his work. The ledger at the asylum reads, "Mr. Carnot Lazare Sadi, ex-military engineer, admitted 3 August 1832, for mania. Cured of mania. Dead from cholera, 21 August 1832."

He was thirty-six years old.

# CHAPTER THREE

# The Creator's Fiat

I have neither propelled vessels, carriages, nor printing presses.
My object has been, first to discover correct principles.
—James Joule

On May 24, 1842, two brothers in their twenties rowed out to the middle of Lake Windermere, the largest in England's Lake District. While the older worked the oars, the younger, who sat with a slight but noticeable stoop, was busy cramming a pistol with gunpowder. The purpose? To study echoes by listening to gunshots as they reverberated around the hills. To ensure a loud bang, the young man, whose name was James Joule, crammed three times the normal amount of gunpowder into the gun. The recoil sent the weapon flying into the lake—one example of Joule's lifelong zeal for scientific experimentation, which, in its early days, showed little regard for health and safety. On another occasion, a misfire blew off his eyebrows. On still another, he gave himself and friends electric shocks, and his cruelest experiment was to use a powerful battery to shock a servant girl while asking her to describe how she felt. Joule increased the voltage until she fainted.

James Prescott Joule, the second of a brewer's five children, was born in 1818 in Salford in Lancashire. Some forty years earlier, Richard Arkwright had installed the world's first steam-driven cotton mill in nearby Manchester, when it was an unremarkable market town in the northwest of England. In the intervening years dozens of new mills had appeared as industrialists pioneered the factory system of mass production in Manchester's increasingly overcrowded streets. Between 1801 and 1830, the population roughly doubled to around 140,000, as people flocked in from all over the country to toil in a city that became known as Cottonopolis.

The Joules, being brewers, prospered. The city's new workers were thirsty, and demand for beer soared to the extent that soon after Joule was born, his father could afford a large house with six servants in the well-to-do neighborhood of Swinton.

Joule, by his own account, was a sickly child—he had regular treatment till the age of twenty for a spinal weakness, which left him with a slightly hunched back. A shy boy, he was deeply attached to his older brother—their parents decided to educate the pair at home so they wouldn't be separated. The family's wealth meant that when Joule was sixteen, his father enrolled him for private lessons with the famous chemist John Dalton.

Joule started work in the family brewery while in his teens and would play an active role in running the business for nearly two decades. In the early years, he attended the brewery daily from 9:00 a.m. to 6:00 p.m. The machinery he encountered there—pumps and vats in which liquids were stirred and heated to precise temperatures—determined the direction of Joule's scientific research. These devices set him on a scientific collision course with Sadi Carnot.

For Carnot, obsessed as he was with steam engines, the key question was how to produce the greatest amount of motive power from a given amount of heat. Joule's surroundings encouraged him to go further, to ask if there might be a better source of motive power than heat. The family brewery did employ a steam engine, and Joule knew how much his business spent on coal. So, with an eye to the bottom line mixed in with a good measure of scientific curiosity, he wondered if a recent invention, the battery-powered electric motor, could drive the brewery's pumps and stirrers more cheaply than burning coal.

Electric motors had been invented in the early 1830s, and within a few years they had become a craze. "Electrical Euphoria" swept the Western world. Societies such as the London Electrical Society convened, Russia's czar and the US government both funded research to see if these new devices could propel boats and pull trains. In Manchester, a magazine called *The Annals of Electricity* appeared. Its editor was a friend of the Joule family, and this obscure periodical would publish much of Joule's early work.

By 1840, ensconced in a laboratory he'd set up in the family home, Joule was building batteries, electromagnets, and motors, and investigating their behavior. One of his earliest observations was his most significant. He noticed that as an electric current runs through a wire, the wire becomes warmer. Electricity, in other words, can produce heat as well as do work by turning a motor. (From now on I shall use the word *work* to mean what Carnot called motive power [i.e., it's a measure of the effort needed to raise a known weight by a known height].)

Electricity's ability to produce heat fueled Joule's suspicion that something was awry with caloric theory, which stated that heat could neither be created nor destroyed. To Joule's eyes, it seemed as if electricity was indeed creating heat as it ran along a wire.

Joule measured this effect with his hallmark diligence and deduced that whether or not caloric theory was true, there is a mathematical relationship between the heat produced, the magnitude of the current, and the resistance of the wire through which it flows. Joule was convinced this was an important discovery, deserving of a wider audience than the readers of the *Annals*, so he sent a paper describing it to Britain's most prestigious scientific publication, *The Transactions of the Royal Society*. Though the equation Joule derived is now taught as part of high school physics and is the basis of every electric toaster, the editor rejected it, allowing only a brief summary to be printed in a lowlier sister journal. It was the first of a series of setbacks Joule would face trying to inform the wider scientific community of his efforts.

Through 1840 and 1841, Joule was becoming ever more skilled at working with electricity, and he focused on comparing the cost of extracting work from an electric motor with that of obtaining it from a steam engine. In Joule's time, batteries consisted of zinc plates that were suspended in a bath of acid. As the zinc dissolved into the acid, electricity was produced, which could power a motor to raise a weight—i.e., do work. In his experimental setup, Joule calculated that the electricity produced as a pound of zinc dissolved could lift a weight of 331,400 pounds by a height of one foot. From a cost perspective, this compared very unfavorably with coal-fired steam engines. In these, burning a pound of coal, a much cheaper material than zinc, could lift a weight five times greater, 1.5 million pounds, to a height of one foot.

This finding killed any idea of replacing the steam engine in his family brewery with an electric motor—"I almost despair of the success of electro-magnetic attractions as an economic source of power," he wrote. But, importantly, it did teach Joule that different ways of producing work are numerically comparable.

Next, Joule started experimenting with dynamos—machines that turn work into electricity. Dynamos, such as the ones attached to bicycle wheels, consist of a wire coil surrounding a magnet. As you pedal, the wheel makes the magnet spin, and this induces an electric current in the coil, which then powers the lights. Joule observed that the electricity produced by a dynamo warms a wire just as that from a battery does. Already suspicious of caloric theory, he now felt he had a way to test it.

To Joule, the ability of an electric current to produce heat had two explanations:

1. Heat was, as most scientists believed, caloric. In which case, to heat up the wires connected to it, the dynamo must be pumping caloric from somewhere within itself to the wires that were warming up. But if that was the case, one would expect the coil within the dynamo to cool down as caloric flowed out of it to the rest of the circuit.

2. Electric current was being *converted* into heat as it flowed through the wires.

So, in late 1842 and early 1843, Joule carried out a series of groundbreaking experiments to determine which explanation was true. He designed a hand-cranked dynamo with a clever modification. He placed the coil in which the electric current is induced inside a glass tube. Joule then filled this with water so he could detect any temperature changes that occurred in the coil. If caloric was real, as Joule turned his dynamo and created electricity, caloric should flow out of the coil, thus cooling the water that enclosed it.

The opposite happened. Far from cooling down, the coil warmed up. Moreover, the more vigorously Joule cranked his dynamo, and the more electric current he produced, the hotter the water in which the coil was submerged became. It seemed that the electricity flowing from and through his dynamo was creating heat and not that the dynamo was moving caloric from one place to another.

To study this further, Joule connected a battery to his dynamo. Before the latter was turned on, electricity flowed from the battery through the dynamo's wire coil and warmed it. That was to be expected—Joule had long ago noted that current from a battery heats a wire. More significant was what happened when Joule cranked the dynamo while it was still connected to the battery. If he rotated the dynamo in one direction, say clockwise, he found the current it generated added to the current flowing from the battery and the water temperature surrounding the coil rose by a greater amount than when the dynamo was not turning. He then cranked the dynamo in the opposite direction, counterclockwise, so the current it generated opposed the current coming from the battery. Now he found that though the water temperature still rose, it did so by a smaller amount. It was as if the electric current from the dynamo was erasing some of the heat that the current from the battery produced.

To Joule, the implications were clear. He wrote with unequivocal confidence, "We have therefore in magneto-electricity an agent capable by simple mechanical means of destroying or generating heat."

The experimental rig Joule had created seemed to him to work in a two-step process. First the work done turning the dynamo generated electricity, and then, as that electricity flowed, it created heat. This implied that the ultimate source of the heat in this arrangement was the work, with the electricity acting as an intermediary.

Joule's next step was to try to quantify this process. If work can be turned into heat, how much of it is needed to create a given quantity of heat? To Joule, work and heat had become interconvertible, the one into the other much like the dollar and the pound. They are both forms of money, and if one knows the exchange rate, one knows how many dollars are equivalent to one pound. Joule believed there was an "exchange rate" between work and heat. Naming this "the Mechanical Equivalent of Heat," he set out to discover its value.

To do so, Joule connected his dynamo to a falling weight via a system of ropes and pulleys. As the weight fell, it turned the dynamo, generating first electricity and then heat, which as before warmed a tube of water. Joule could now equate the height by which a known weight fell to the amount of heat that was created. In other words, he could measure the Mechanical Equivalent of Heat.

Joule defined what he called a "unit of heat" as the amount of heat needed to raise the temperature of a pound of water by 1°F. Joule further declared a unit of work to be the amount provided as a one-pound weight falls by one foot, which he referred to as a foot-pound. Over several weeks Joule ran versions of this experiment over and over with painstaking meticulousness. The process was fiddly and difficult, not least because the heat from the electric current raised the water temperature in the tube by, at most, 3°F, a figure barely discernible on his thermometer. In addition, Joule struggled to insulate his apparatus from the ambient temperature of his room, so that he could ensure any temperature changes were solely due to the electric current generated in his dynamo.

After several weeks of experimental labor, Joule satisfied himself that his results were reliable. Better yet, those results showed there did indeed seem to be a fixed conversion rate between work and heat. The value was hard to pin down exactly—it seemed to lie somewhere between 750 and 1,000 foot-pounds per unit of heat—and so Joule took an average of all his readings.

"The quantity of heat capable of increasing the temperature of a pound of water by one degree of Fahrenheit's scale is equal to, and may be converted into, a mechanical force capable of raising 838 lb. to the perpendicular height of one foot."

To modern eyes Joule may seem to have been a little too eager to believe his results. His confidence stemmed in part from his upbringing and beliefs. Politically conservative and a devout Christian, Joule saw his scientific endeavors as "essentially a holy undertaking," convinced as he was that a divine being had endowed the universe with a fixed amount of an immaterial substance that enabled change and movement. Electricity, work, and heat were simply different facets of this. They might be converted from one to the other, but the total amount was invariable. As Joule wrote toward the end of one of his papers, "The grand agents of nature are, by the Creator's fiat, indestructible; and that wherever mechanical force is expended, an exact equivalent of heat is always obtained."

What Joule called *the grand agents of nature* we today know as *energy*. For what lay behind Joule's religious words was the elaboration of a principle known to scientists today as the conservation of energy and also as the first law of thermodynamics.

In the summer of 1843, hoping to promote his discovery, Joule traveled to Cork, on the south coast of Ireland, to attend a meeting of the British Association for the Advancement of Science. The BAAS had been founded a decade earlier by British scientists frustrated by the elitist and conservative character of the Royal Society. The word *scientist* was mooted at their first meetings. The purpose of the term was to unite under a single banner "students of the knowledge of the material world." The BAAS rated Joule's work highly enough to invite him to speak about the Mechanical Equivalent of Heat, but, as he later reported, "the subject did not excite much general attention." Why? One factor was that Joule was an outsider finding fault with caloric theory, which many felt was well established. Also, he was not charismatic. His dress, "commonplace in the extreme," his manner "nervous," "he possessed no natural grace" and "no great facility of speech." Even while presenting evidence for one of the most important ideas in science, Joule couldn't cut through. The following year he submitted yet another paper to the Royal Society. Again, they declined to publish.

Undeterred, Joule worked on, determined to make the case for the inter-convertibility of heat and work unassailable. His next step was his most famous experiment and, conceptually, the simplest. For having demonstrated that work could be turned into heat with electricity as an intermediary, Joule now wanted to show that the electrical stage wasn't essential, and that work could be converted directly into heat. To do this he drew on the widely observed fact that the friction between any two objects as they're rubbed together generates heat. Show that this process turns work into heat at the same "exchange rate" as he had measured in his dynamo experiments, and he would bolster his case.

For this experiment, Joule was inspired by a type of machine that would have been familiar to a brewer—a vat in which a liquid mulch of hops and barley is stirred by a mechanical arm as part of fermentation. Joule's apparatus scaled this down. He made a metal cylinder about a foot tall and about eight inches in diameter. He then filled it with water and placed inside it what he called a "paddle wheel," a mechanism to stir the water as it turned. Joule's belief was that the frictional forces between the vanes of the paddle wheel and the water they moved through would create heat and raise the water's temperature. Then, as with his dynamo experiment, he attached the axle of the paddle wheel, via a system of ropes and pulleys, to a falling weight. Thus, as the weight fell, it turned the paddle wheel, which in turn warmed the water. He could then equate any rise in the water temperature with the work done as a known weight fell by a known height.

The experimental challenge here was how best to accurately measure the temperature rise in the water. Joule's thermometers consisted of a column of mercury trapped in a thin glass tube, which expanded and contracted as it was heated and cooled. Early on in his experiments, Joule realized that even vigorous rotations of the paddle wheel made the water temperature rise by less than 1°F. Though this was enough to cause the mercury to expand, it was only just visible to the naked eye and was far too small to verify the value of the Mechanical Equivalent of Heat.

Fortunately, Joule had in the mid-1840s encountered a Manchester lens and spectacle maker, John Benjamin Dancer, who had found fame pioneering a technology known as microphotography. This involved printing tiny images of about a square millimeter in size on pieces of photographic paper. Dancer constructed microscopes through which the public were invited to view these microphotographs, whereupon the original image was rendered visible. Though microphotography had no scientific purpose—the images ranged from the Ten Commandments to St Paul's Cathedral—Joule recognized the usefulness to him of making small things visible. So he commissioned Dancer to design a thermometer, which with the aid of a lens, allowed minute movements in a thin column of mercury to be magnified and measured accurately.

Dancer's thermometers, Joule claimed, enabled him to measure temperatures to an accuracy of better than one-tenth of a degree Fahrenheit. When used with the paddle-wheel apparatus, their readings supported the conclusions from Joule's dynamo experiments. In those, he had estimated the mechanical equivalent of heat as 831 foot-pounds. The new experiment gave the figure as 781.5 foot-pounds. Joule repeated the experiment with sperm-whale oil in the cylinder and obtained a similar result—781.8 foot-pounds.

These experiments carried out in different conditions and with different materials all pointed to a value for the mechanical equivalent of heat of a little below 800 foot-pounds. With his confidence in the validity of the concept thus boosted, Joule tried again to be heard. On April 28, 1847, he gave a lecture to the general public in a church in Manchester, where he presented his evidence and repeated his claim that energy conservation was part of a divine plan. A local paper, the *Manchester Courier*, published the lecture in full, and Joule sent copies to his friends. The scientific community, however, remained uninterested.

That summer, the BAAS annual meeting was in Oxford, to which Joule was again invited. He had no expectations of a favorable reception, and to make matters worse, there was a scheduling foul-up on the day of his talk. The organizers told Joule, who was due to speak to a group of chemists, that they were now too busy for his lecture. So, instead, they asked him to speak to some physicists, who had time available. The organizers made it clear to Joule that he should make his talk as brief as possible.

For Joule and for science, however, this organizational snafu was serendipitous. Finally, after a decade of scientific obscurity, chance had placed someone in the audience who seemed interested in Joule's work. At the end of the talk, a young man stood up and began asking questions, creating "a lively interest in the new theory." The interrogator was a twenty-three-year-old from Glasgow named William Thomson, who was already regarded as one of his nation's leading scientific minds. Years later he, too, vividly recalled the encounter and the feelings of both intrigue and alarm it had evoked: "I felt strongly impelled at first to rise and say that Joule must be wrong," but "as I listened on and on, I saw that Joule had certainly a great truth and a great discovery, and a most important measurement to bring forward."

At the time, Joule's "great discovery" presented Thomson with an uncomfortable dilemma. Over the previous two years, he'd fallen in love with Sadi Carnot's elegant analysis of the way an unchanging amount of caloric produces work as it flows from a hot furnace to a cold sink. Yet, here was this unassuming Mancunian claiming caloric did not exist. The easy option, taken by many others in the room, would have been to dismiss Joule's evidence—much of it was based, after all, on minuscule temperature increases discernible only on novel thermometers.

William Thomson, however, possessed remarkable scientific intuition. In his mind, Carnot's theory and Joule's experiments both rang true despite appearing incompatible. Could they both be right? If so, how?

Decades later, William Thomson, ennobled as Lord Kelvin for his contributions to science, embellished the story of the Oxford encounter while unveiling a statue of Joule. According to Thomson, while on a walking holiday a few weeks after that first meeting, he'd run into Joule near the Alpine resort town of Chamonix. Joule was on his honeymoon, but he'd left his bride in a nearby carriage while, thermometer in hand, he searched for a waterfall to confirm a theory that the temperature at the top is cooler than at the bottom. A fortnight later, Joule was still on this quest, and Thomson even joined him in attempting to measure temperature differences at the Cascades de Sallanches. Thomson may, however, have invented this anecdote to convey Joule's unwavering commitment to science. In a letter to his father written shortly after the Alpine meeting, he makes no reference to thermometers or waterfall temperatures. But when Thomson recounted the tale in later life, he described it as "one of the most valuable recollections of my life."

CHAPTER FOUR

# The Valley of the Clyde

*Caino? Je ne connais pas cet auteur.*
—Paris bookseller to William Thomson

In 1845, two years before the encounter in Oxford, while Joule was laboring in his home laboratory in Manchester, William Thomson was traipsing the streets of Paris from bookshop to bookshop searching for Sadi Carnot's treatise, *Reflections on the Motive Power of Fire*. Thomson had come across a description of *Reflections* in a French scientific journal, and what he'd read had fired his imagination. It represented a breakthrough, he became convinced, in the understanding of heat. But because of the inquirer's Scottish-accented French, the booksellers struggled to recognize the name of the author whose book he was seeking. Even when he managed to clarify it by overemphasizing the *r* in Carnot, he was offered a book on "some social question" by Sadi's brother, the politician Hippolyte Carnot.

Thomson was in Paris undergoing the final stages of an education designed from boyhood to prepare him for scientific greatness. Born in Belfast in 1824, he'd moved to Glasgow eight years later with his family when his father had been appointed mathematics professor at the city's university. Aged fifteen, Thomson won the class prize at the same institution for an analysis of how the earth's shape had formed. A year later his precocious mathematical talent further manifested when he encountered *The Analytical Theory of Heat* by the French polymath Joseph Fourier. This text was striking in that it made no claim as to what heat is. Fourier's aim was to mathematically describe how heat behaves and, in particular, how it flows. An example is a metal bar that is hot at one end and cold at the other. Experience tells us that heat will diffuse from hot to cold until the bar's temperature equalizes. Fourier showed how this kind of diffusion can be described mathematically. His approach was unusual for the time, and Fourier had critics. Aged sixteen, Thomson published a detailed defense of the Frenchman's methods in the scholarly *Cambridge Mathematics Journal.*

Proudly aware of Thomson's talent, his father urged him to study mathematics at Cambridge. Over the previous two decades, a new generation of professors there had embraced the latest developments in the subject from Europe, and the university had regained its reputation as the nation's preeminent center for mathematical training. At Cambridge, Thomson's peers and teachers confirmed his promise. Senior academics who read his essays were surprised to learn the author was still a teenager.

Meanwhile, Thomson's father began hatching a plan for him to become Glasgow University's next professor of natural philosophy. (Later in the nineteenth century, this phrase was replaced with the modern term *physicist.*) With the incumbent old and in poor health, Thomson senior's one concern was that the post would not be open to a youthful candidate whose only credentials were a Cambridge mathematics degree—a qualification that showed a talent for abstract reasoning but not necessarily for demonstrating physical phenomena to students, a highly valued skill at Glasgow University, the leading educational establishment in an industrial city that prized practicality. Thomson senior had heard, however, that France led the way when it came to teaching science by demonstration. He urged his son to obtain letters of introduction to eminent French savants, travel to Paris, and get his hands dirty once he'd received his Cambridge degree.

In Paris, Thomson worked as an assistant to an experimental physicist, Victor Regnault, who had been funded by the French government to study the thermal properties of steam. (Regnault, like Carnot, had benefited from the French revolutionary government's commitment to widespread education. Orphaned at eight and impoverished, he had won a place at the Polytechnic School and had gone on to become one of France's leading scientists.) Working in Regnault's laboratory, even in a lowly capacity, holding test tubes or operating an air pump, was a transformative experience for the young Thomson. He observed firsthand how water and steam behave when heated and cooled. Shadowing the diligent Regnault also taught

Thomson the importance of patience and precision in experimental physics. And crucially, it was in Paris—"without doubt the Alma Mater of my scientific youth"—where Thomson encountered the ideas of Sadi Carnot.

In April 1845, Thomson returned to Britain, where, as his ailing predecessor clung to life, he had to wait another year for the Glasgow professorship to become available. In the interim, Thomson made ends meet by coaching Cambridge undergraduates and continued to mull what he had gleaned of Carnot's ideas. He discussed them in detail with his older brother by two years, James, who was a gifted scientist in his own right. Though always coming second to William, he'd excelled as a student at Glasgow University, and then worked as an apprentice to several firms in England and Scotland. James had a passion for engineering—"He talks about it ceaselessly all day"—and he and William were a formidable scientific double act. William had mathematical skills and a good grasp of laboratory physics. James had hands-on experience working with real steam engines. William's mind was sharp and mobile, James's dogged and stubborn. The pair loved nothing better than to discuss science and engineering, although listening in wasn't easy: "It is really quite comic to see how both brothers talk at one another, and neither listens."

To the Thomson brothers, Carnot's analysis rang true. Its abstract reasoning appealed to William, and James saw it in action, particularly because of the time he'd spent working on steam engines intended for ships. These forced their builders to focus on efficiency, even if coal was cheap. The weight of the fuel limited how much a ship could carry and how far it could go. Intuitively, James felt it was wasteful that these engines' condensers operated at warmer temperatures than that of the surrounding sea or ocean. (The condenser is where steam is turned back into water, so it no longer presses on the piston.) To James, this meant that the heat in the condenser was being wasted as it wasn't doing any work. If a way could be found to condense steam at the same temperature as the ocean, then, James believed, ships could go farther without consuming more coal. This view chimed with Carnot's analysis—"a very beautiful piece of reasoning" is how James described it to William.

In September 1846, the professor of natural philosophy at Glasgow died, and William Thomson, then only twenty-two, took up his post. Here he set up Britain's first physics laboratory in which undergraduate students took part in laboratory work. Thomson was an enthusiastic teacher who got mixed reviews from students—his passion for science inspired, his tendency to digress confused.

Thomson made investigating the properties of heat and steam a priority, an inevitability given that just outside the university's walls, just as in Manchester, industrial activity had reached unprecedented levels. Glasgow's focus was on ships rather than textiles, and several of Thomson's students were the sons of shipbuilding and engineering families. During that decade, a new vessel slid out of the yards along Glasgow's river Clyde every ten days. One of the most magnificent was the 1609-ton S.S. *City of Glasgow*, which boasted an iron hull and a propeller rather than a paddle wheel and could take over four hundred passengers across the Atlantic to America's east coast in under three weeks. Such vessels helped populate America.

Alongside shipbuilding, Glasgow saw a surge in ancillary and associated industries, everything from cotton mills to chemical factories and iron foundries. Immigrants from Ireland and the Scottish Highlands flooded into the city to find work, and Glasgow's population rose from seventy-seven thousand in 1800 to around three hundred thousand in 1850. A visitor who came to stay with his brother in the city in 1851 reported that he knew that "six o'clock had struck by hearing, far down below him in the Valley of the Clyde, the thud of a great steam hammer, to which a thousand hammers, ringing on a thousand anvils, at once replied, telling that the city had awakened to another day of labour."

To William Thomson, Sadi Carnot's ideas underpinned the sound by which his city marked time. His education, his discussions with his brother, and his intuition all pointed to that conclusion. This explains his mixed emotions on encountering James Joule in Oxford in 1847. Joule's conviction that heat and work could be turned into each other flew in the face of a key assumption that Carnot had made, namely that heat, being caloric, could not be created or destroyed.

Then, in the autumn of 1848, William Thomson obtained a copy of Carnot's original treatise. Whatever doubts Joule had seeded in his mind, Thomson was now convinced the Frenchman's work was too important to remain in scientific obscurity. He set to work writing a paper designed to disseminate its ideas and present evidence in their favor. Its title, "An Account of Carnot's Theory of the Motive Power of Heat with Numerical Results Deduced from Regnault's Experiments on Steam," undersells its importance. It is not merely an English rendition of the French original; rather, the paper is one of the most significant Thomson wrote over a long and illustrious career, not least because with it he introduced a new word into the scientific lexicon: *thermodynamic*.

More significant is the seemingly contradictory strategy Thomson employed in his "Account." The main body goes to great lengths to demonstrate both theoretical and experimental support for Carnot's treatise. But at regular intervals, Thomson introduces footnotes referring to Joule's work that serve as troublesome critiques. The paper captures the conflict between Carnot and Joule in Thomson's mind. Though he can't resolve it, by juxtaposing each side of this internal debate in the same text, he would enable others to join in.

In 1849, Thomson published his paper on Carnot, doubts and all. Then unexpectedly, within a few months, new evidence emerged that, on the face of it, favored Carnot over Joule.

This evidence came from James, William's older brother, who had devised an ingenious way of testing Carnot's ideas. James wondered if he could come up with an engine design that might contradict Carnot's hypothesis that there must be a fall in temperature in the working material of the engine for it to produce motive power. If such a counterexample existed to a principle that Carnot claimed was universal, the entire theory would be in jeopardy. But what if the counterexample actually substantiated the theory?

James Thomson knew that any substance that expands can, in principle, push a piston and thus create work. He also knew that water expands on freezing. So, why not design an engine that exploited this fact? Thomson imagined a piston/cylinder arrangement like the one in a steam engine in which the space below the piston is filled with water. This is cooled to 0°C. When the water reaches this temperature, and only then, it turns to ice, expands, and pushes the piston. The crucial aspect of this arrangement is that the piston is pushed at an *unchanging temperature of zero degrees*. Work has been produced, it seems, with no fall in temperature. Is this the counterexample capable of demolishing Carnot's theory?

Not necessarily. For as the water freezes, expands, and pushes at the piston, the piston pushes back at the ice. This is a consequence of the idea that for every action there is an equal and opposite reaction. Imagine you were shrunk down inside the cylinder and were pressing with your hands on the piston. You would feel a resisting pressure from the piston on your hands. In the same way in James Thomson's hypothetical ice engine, as the water freezes, the pressure on it from the piston would increase.

In Thomson's time, water was known to freeze at 0°C on the earth's surface, where the only pressure acting on it is the weight of our planet's atmosphere. (Scientists refer to this as a pressure of "one atmosphere.") But it was not known whether water's freezing point would change if subjected to higher pressures. If these caused ice to form below 0°C, then Carnot's theory would be saved. Because in that case, as the water turns to ice and produces work, the temperature drops from 0°C to whatever the freezing point of water is when it's under pressure from the piston.

James Thomson calculated the amount by which the freezing point of water must drop as the pressure on it increases for the theory to hold. He worked out that for every increase of pressure by one atmosphere, the freezing point of water should drop by 0.0075°C. So, at a pressure of two atmospheres, water should freeze at minus 0.0075°; and at a pressure of three atmospheres, it should freeze at minus 0.0150°, and so on.

On reading these calculations, William Thomson was elated. He now had a way of testing Carnot's theory in his laboratory at Glasgow University. If he could measure the drop in the freezing point of water as it's put under pressure to be a value predicted by his brother, it would provide evidence that Carnot's theory was correct.

Of course, such an experiment presented huge practical difficulties, not least because the predicted fall in temperature predicted was tiny— too tiny for thermometers of the time to detect. So, in late 1849, William Thomson commissioned one of his students, Robert Mansell, to build a thermometer sensitive enough to detect temperature changes of less than one one-hundredth of a degree Celsius. Mansell had had training in practical engineering prior to enrolling at Glasgow University, and he also had knowledge of glassworking. He took painstaking efforts to calibrate the thermometer to the point where Thomson felt its readings were reliable. Then Thomson filled a glass cylinder with water, which he could press down on with a piston. With this arrangement, Thomson measured the temperature at which water froze under differing amounts of pressure.

Much to William Thomson's satisfaction, the results vindicated his brother and, by extension, Carnot. James Thomson's theory predicted that under a pressure 8.1 times greater than atmospheric pressure at sea level, the freezing point of water would be minus 0.061°C. William Thomson measured it as minus 0.059, a tiny discrepancy. Similarly, at a pressure 16.8 times greater than the atmospheric pressure at sea level, the theory predicted the freezing point of water would be minus 0.126°C. Thomson measured it as minus 0.129, again a minute discrepancy.

For Thomson, the ice experiment was compelling evidence in favor of Carnot's theory. Yet, as with Joule, from a twenty-first-century perspective, Thomson perhaps appears too easily convinced. Two data points, which Thomson admitted could be the result of chance, prove little. That they satisfied him was as much emotional as it was logical. He was in love with Carnot's elegant analysis of how heat produces motive power, and he saw his experimental results as confirmation of what he felt in his heart must be true.

The Thomson brothers had also, without knowing it, explained how glaciers move. The pressure on the ice at the bottom of a glacier is so great that it melts, even though the temperature there is 0°C or lower. A layer of water is thus created under the glacier, allowing it to slide downhill. Remarkably, Sadi Carnot's reflections on how steam engines work had led to an explanation for the movement of glaciers.

Thomson knew, however, that though the ice experiment was evidence for Carnot, it was not evidence against Joule. The latter's arguments against caloric theory still held. Thomson could not simply dismiss them. In fact, Joule's work amplified another doubt Thomson had about the theory, which was this:

Though the flow of heat from hot to cold can produce work, this flow doesn't always do so. Take, for example, an iron bar that's red-hot at one end and cool at the other. Over time, heat will flow across the bar, from the hot to the cold end, equalizing its temperature. What happens, wondered Thomson, to the work this flow could have produced? If heat is an indestructible fluid, caloric, an iron bar with a temperature differential is analogous to a bucket of water at the top of an inclined channel. Tip the bucket and water will flow down the channel just as in the iron bar heat will flow from the hot to the cold end. But imagine that a paddle wheel is halfway down the water channel. As the water flows past it, it turns and raises a weight. As some of the motion of the water is imparted to the wheel, its speed of flow decreases. When the water reaches the end of the channel, it makes a splashing sound. Remove the paddle wheel and the water rushes down unimpeded, smacking the end of the channel with a much noisier splash. The power the water could have imparted to the paddle wheel has ended up as sound. And here lay a problem for Thomson. As caloric flows unimpeded down an iron bar from the hot to cold end, it doesn't make a splashing sound or anything equivalent. So what happens to the power it could have produced? Thomson had no answer.

For his part, Joule had realized that for all the flaws in caloric theory, Carnot's conclusion that heat can only generate work as it flows from a hot furnace to a cool sink could not be dismissed. In March 1850, he wrote to Thomson saying, "There ought to be some connecting link between the results I have arrived at and those deduced from Carnot's theory. Perhaps you will succeed before long in discovering it. For my own part it quite baffles me."

Carnot and Joule were two pieces of the puzzle. But despite their best efforts, neither Thomson nor anyone else could fathom how they fit together. For that to happen, the talents and efforts of an ambitious and emergent nation would have to be engaged.

# CHAPTER FIVE

# The Principal Problem of Physics

It is a spectacle for the gods to see the muscle working like the cylinder of a steam engine.

—The Berlin-based physiologist
Emil du Bois-Reymond

In Berlin's southwest corner, where the German capital borders the town of Potsdam, the river Havel opens out into an interconnected system of lakes, canals, and waterways. On their shores lie a collection of parks, gardens, and palaces, which in the first half of the nineteenth century were the playgrounds of the Hohenzollern royal family. At the time, they ruled what was then the kingdom of Prussia, a territory that corresponds roughly to the northeastern quarter of modern Germany.

One of these pleasure gardens, the Glienicke Park, resembled an English landscape garden with fountains, greenhouses, and extensive flower beds. One visitor, Helmuth von Moltke, described it as "one of the most beautiful in Germany."

Walk in von Moltke's footsteps today and you see much the same sights. But some are missing, notably a footbridge, which was intended to appear ruined and beneath which flowed a powerful stream of water. A few hundred yards past it, one would have heard a strange sound, now also long gone—an insistent clattering and puffing, emanating from within a villa that would not have seemed out of place in Renaissance Florence. Entering, one would have seen a steam engine, one of the earliest in Prussia and designed by an engineer trained in England.

Von Moltke wrote of this machine that it "works from morning till night to lift water out of the Havel up to the sandy heights to create lush meadows where without the engine only heather would survive. A pow-erful cascade roars over the cliffs under the arch of a bridge, half-washed away, seemingly from its violence, and abruptly rages fifty feet down to the Havel onto a terrain where prudent Mother Nature would not have thought to let a pail of water flow." In other words, this was an artificial landscape, made more beautiful by machinery.

In Britain, steam power was seen as the route to commercial profit, in France to social progress, and in Prussia, in elite circles at least, to improve on nature. Here steam power was connected to the natural world in a literal way, and here, scientists first appreciated that the lessons learned from steam power applied far beyond the engines themselves. After all, if a steam engine could enhance nature, perhaps it could also explain it? One young man, who had witnessed the installation of the Glienicke Park's steam engine and that of two others in nearby gardens, was one of the first scientists to see this larger relevance.

Born in 1821 in Potsdam, Hermann Helmholtz came from a middle-class family—his father was a teacher at a gymnasium, a Prussian second-ary school that emphasized academic rather than practical education. By his own account, Helmholtz was a sickly child who spent long periods in bed or confined to his room. As he grew older and stronger, however, his father introduced him to a wide range of literature and poetry and took him to Potsdam's parks and gardens. In tandem, Helmholtz devel-oped an early fascination with mathematics and science, which he taught himself by devouring physics textbooks and building homemade micro-scopes from disused spectacle lenses. As he entered his late teens, Helm-holtz's intellectual self-confidence grew, and in 1838 he won a scholarship to study medicine at the Friedrich Wilhelm Institute in Berlin, an orga-nization set up primarily to train army surgeons. The year 1838 also saw the establishment of the first steam-powered train line in Prussia, running from Potsdam to Berlin. So not only did the young Helmholtz witness the nature-enhancing uses of steam power in his local park, but he now expe-rienced its more utilitarian function as a method of transport that took him from home to university.

In the late 1830s and early 1840s, the German-speaking part of Europe was a patchwork of kingdoms, archduchies, bishoprics, principalities, and other sovereign territories. Economically it lagged behind both Britain and France as it did with steam technology. In 1840, the capacity of fixed steam engines in the region's factories was a mere 20,000 horsepower, far behind the 350,000 horsepower in Britain and the 34,000 in France.

As the 1840s unfolded, however, reforms enacted in earlier decades began to pay off. Prussia had abolished serfdom in 1807, allowing peasants to live and work where they chose, which created a numerous and mobile workforce. In addition, a coalition of German states had formed a customs union in 1834. Previously, a journey from Hamburg in the north to the Alps in the south would have necessitated crossing ten states and at each border dealing with "hostile tax-gatherers and customs house officials." Now that the system was abolished, textiles, mining, and steel production all began to accelerate. Between 1840 and 1860, fixed steam engine capacity increased tenfold and by 1869, the railway network had expanded to over ten thousand miles of track.

In tandem, a much-reformed and better-funded German education system evolved. Over the first half of the nineteenth century, the government of Prussia increased spending on her universities fivefold and redefined their purpose. No longer were they institutions where students were spoon-fed existing knowledge; instead they were places in which new learning was to be acquired. Erudition wasn't sufficient by itself; academics were expected to carry out original research and to demonstrate creative thought.

This was the society in which the young Hermann Helmholtz grew up. He took full advantage, forming long-lasting friendships with other ambitious young physicians, physicists, and chemists. A like-minded group coalesced with the common mission of bringing the study of living organisms in line with existing research into the inanimate world. In modern terms, they wanted to show that living organisms obeyed the same mathematical, physical, and chemical laws as everything else. However, this approach put Helmholtz and his network in conflict with a large section of the European scientific community who felt such a synthesis of the animate and inanimate worlds was not possible. Many scientists of the day believed in vitalism, the idea that living organisms, in addition to the sustenance they received from food, water, air, and so on, also possessed a "vital," life-giving force. While an organism was alive, this vital force controlled the physical and chemical processes that took place within it. Logically, therefore, when it died, that vital force disappeared, leaving the dead organism to decay as if it were inanimate. Helmholtz and his friends opposed this "vitalist" view and felt disproving it was a crucial step to putting biology on the same footing as physics and chemistry.

In 1843, Helmholtz graduated from medical school and took up his post as assistant surgeon to a regiment of Red Hussars in his hometown of Potsdam. Despite military duties that began at 5:00 a.m., when a bugler blew reveille, Helmholtz remained committed to his scientific mission. He set up a small laboratory in his barracks at his own expense and began a series of experiments to investigate the validity of vitalism. Of particular interest to Helmholtz were recent studies into the sources of animal heat, which had become a battleground.

The feeling in the anti-vitalist camp was that if they could demonstrate that the way warm-blooded creatures generate their warmth is akin to a slowed-down form of burning, not different in principle to the combustion of coal, a telling blow would be struck against vitalism.

This hypothesis had been promoted as far back as the 1780s, by the great French chemist Antoine Lavoisier, who pictured the lungs as a torpid fireplace where food burned: "Respiration is then a combustion, admittedly very slow, but nevertheless completely analogous to that of charcoal." Food, in other words, was fuel that animals burned in oxygen and, in so doing, generated heat with carbon dioxide as the waste product.

In the following years, however, scientists realized the process is rather more complicated. Foods are not like charcoal, which is made up entirely of the element carbon. Sugars and other carbohydrates, for example, are complex molecules that contain atoms of hydrogen and oxygen as well as carbon. So, in addition to the contribution to animal heat made by burning carbon, there could also be a contribution from burning hydrogen. This also releases heat, with water ($H_2O$) as the end product. The observation that animals exhale carbon dioxide and excrete water supports this assumption.

With this in mind, two scientists, the Belgian César-Mansuète Despretz and the French Pierre Louis Dulong, investigated Lavoisier's assertion that respiration is a form of slowed-down combustion. Working separately in the 1820s, they placed rabbits, guinea pigs, pigeons, cocks, owls, magpies, cats, and dogs in copper boxes, which were immersed in a tank of water. This arrangement enabled the scientists to measure how much oxygen an animal inhaled over a known length of time. They then estimated the proportion of the inhaled oxygen that combined with carbon to form carbon dioxide, and the proportion that combined with hydrogen to form water. After that, they measured the temperature increase of the water in the tank. Next, they burned carbon and hydrogen with oxygen to produce the same amount of carbon dioxide and water that the animal had exhaled and excreted. Finally, they measured how much heat this released.

Both scientists observed that simply burning carbon and hydrogen with oxygen released roughly 10 percent less heat than their animals did to generate the same amount of carbon dioxide and water.

This conclusion chimed with vitalist views that there must be another source of heat in animals, one that was not subject to the same laws of physics and chemistry that rule the inanimate world. Helmholtz, following this debate from his army quarters in Potsdam, was skeptical. He decided that he would study animal heat for himself.

Helmholtz attacked Dulong's and Despretz's work with a three-pronged strategy. First, he argued that their experiments were based on flawed assumptions. The pair had measured the heat given off when carbon and hydrogen are burned in oxygen. Helmholtz pointed out that the carbohydrate molecules in food produce more heat when burned than the elements carbon and hydrogen do. This happens because carbohydrate molecules, as well as being made up of carbon and hydrogen, contain a number of oxygen atoms. These are embedded in the carbohydrate molecule. So, when animals respire, they're not just combining carbon and hydrogen with atmospheric oxygen, it's as if they have an additional supply of oxygen that's built into their food. When that's taken into account, the gap between the amount of heat animals release and the amount released by burning disappears.

Secondly, Helmholtz drew on his medical training. He carried out delicate experiments on frogs' legs to try to establish that muscular movement is caused by normal chemical processes and does not derive from a vital force. In simple terms, Helmholtz submerged some frogs' legs in water and then in alcohol and measured the amount of material that leached out of them and dissolved into the liquids. He then passed electric currents through other frogs' legs that had not yet been submerged. This made their muscles twitch. He then placed these electrocuted legs in water and alcohol and measured the amount of material that dissolved out of them. He saw that the muscle movement caused a reduction in the amount of material that dissolved in water, which was exactly balanced by an increase of material that dissolved in alcohol. Muscle movement, in other words, was accompanied by the change of a water-soluble substance into an alcohol-soluble substance. The lesson was clear: muscle movement was driven by chemical energy released as one substance turned into another, again, in principle, no different to combustion.

The third part of Helmholtz's strategy drew its inspiration from the steam engine.

Specifically, Helmholtz employed the same assumption about the impossibility of perpetual motion that Sadi Carnot had made in his seminal paper on steam engine efficiency. If the vitalists were right and animals could generate more heat than could be released by burning carbon with oxygen, then there must be some other source of heat within them that is not subject to physical laws. That, however, would imply that animals could create some heat *without consuming any food or fuel*. And if animals could indeed generate some heat from nothing, that heat could, in principle, be used as a source of work: it could be used to lift a weight, power a fountain, or pull a train and more, without consuming any fuel. From that it followed that if some animal heat came from a nonmaterial "vital" source, it could drive a perpetual motion machine. But, Helmholtz continued, perpetual machines cannot exist. Therefore, animals cannot create heat without fuel. Their body heat must come entirely from the food and oxygen that they consume. Where Carnot had used the impossibility of perpetual motion to show that heat was the ultimate source of work in a steam engine, Helmholtz argued that all the heat released by an animal must come from chemical reactions that obey the same laws as those that govern the inanimate world.

Helmholtz's arguments against vitalism were warmly received by many of his fellow physicians. Emboldened, in early 1847 he embarked on a new paper that he hoped would extend his arguments against vitalism into the whole of science. The key, once again, was the impossibility of perpetual motion. But now Helmholtz thought of this in a radically new way. To this point, the idea that one couldn't obtain work from nothing—pumping water out of a well without some effort or fuel, for instance—was seen as a negative. A price had always to be paid to get anything useful done. Helmholtz's insight was to say that the impossibility of perpetual motion, far from being a bad thing, is an invaluable guide to how the universe works at a fundamental level. It can shed light on the way that disparate phenomena such as gravity, motion, heat, electricity, and magnetism relate to one another. Or as he put it later, "If perpetual motion is impossible, what is the relationship between natural forces that must hold? Everything was gained by this inversion of the question."

In July of 1847, Helmholtz read his paper on this topic at a meeting of the Berlin Physical Society, an organization founded by a network of physicians, chemists, physicists, and engineers that represented the new Prussian technocracy. Entitled "On the Conservation of 'Kraft,'" it's not so much a scientific paper as a manifesto for theoretical physics, written with all the ambition of a twenty-six-year-old brimming with confidence.

Helmholtz's paper did not explicitly state any scientific ideas that others hadn't previously. Its significance was to offer up a guiding principle—the impossibility of perpetual motion—by which all natural phenomena can be viewed and understood. But why is that principle so insightful? The colloquial answer is that there's no such thing as a free lunch. The total amount of what Helmholtz called *Kraft* in the universe is conserved or fixed. Whether heat or electricity or movement, all these forms of *Kraft* can be turned into each other, but nothing is destroyed or created in the conversion. James Joule had come to a similar conclusion, as had a German physician, Julius Robert Mayer from Württemberg. In its broad sweep and its explicit aim of unifying all science under the banner of *Kraft* conservation, Helmholtz's paper was unique.

The hardest aspect of this idea is the word *Kraft*. Directly translated, it means *force* in English. In the context Helmholtz uses it, it's perhaps better translated as *energy*, but energy is not straightforward to define. Even today, most of us are aware that energy resides in gasoline and food and that it flows into our homes as electricity or gas, but we lack an intuitive grasp for why all these disparate phenomena fall under the umbrella term *energy*. This problem also confused many nineteenth-century scientists. Helmholtz gave *energy* clarity by relating it to the impossibility of perpetual motion. For a better sense of why, consider the following thought experiment inspired by Helmholtz's paper:

Imagine a frictionless slope of one meter length at an angle of forty-five degrees to the horizontal. At the top of the slope, ready to slide down it, is a cube of metal that weighs one kilogram. A rope connects the upper side of the weight to an electrical dynamo. The weight slides down the slope till it reaches the floor, and as it does so, it turns the dynamo, which generates electricity. That electricity drives a motor that pulls the weight back up the slope. With this Rube Goldberg arrangement, we have taken energy from the earth's gravitational pull and turned that into the downward movement of the weight. We then turned that movement energy into electrical energy and then back into movement energy, which enabled the weight to go uphill against the pull of the earth's gravity.

Helmholtz's point is that if each of these conversions is done perfectly, so nothing is lost along the way, then the weight will end up exactly where it started. That's the best that can be done. Under no circumstances is it possible to rig these conversions so that the weight will reach a height greater than from where it started.

For Helmholtz, this type of analysis enabled very different-seeming phenomena—gravity, motion, electricity—to be related to each other quantitatively. Every type of energy has a "best possible" exchange rate into another form of energy, which is embedded in the laws of nature.

Helmholtz also introduced another important idea in his paper, which today is called potential energy but which he referred to as *tensional forces*. In simple terms, this means that energy can be stored and released later. So, going back to the arrangement above, when the weight is at the top of the slide, it is effectively a store of gravitational potential energy, which is released as it slides downhill. If, as it turns the dynamo, that electricity is used to charge a battery, that energy is now stored as electrical potential energy, which can be used later to, say, power a motor. Helmholtz pointed out that foods are stores of chemical potential energy, and as animals digest them, they "use up a certain quantity of chemical tensional forces and . . . generate heat and mechanical force." This perspective led him to conclude that the ultimate source of chemical potential energy in food must be sunlight.

Throughout his paper, Helmholtz acknowledges that he doesn't yet know the values for the exchange rates between different forms of energy. But they have to exist, he argues, and one of the aims of physics should be to deduce by experiment and observation what they are. He concludes with a call to action, to "lay before physicists as fully as possible the theoretical, practical, and heuristic importance of this law, the complete corroboration of which must be regarded as one of the principal problems of physics in the immediate future."

Helmholtz's paper is a landmark in the history of physics. At the time, however, it was greeted with skepticism to the extent that Prussia's most prestigious scientific journal, *The Annals of Physics*, turned it down on the grounds that it was too speculative, too theoretical, and lacking in new experimental findings. As with the work of Carnot and Joule, the scientific community seemed underwhelmed by Helmholtz's work. Only with the help of friends in the Berlin Physical Society did Helmholtz find a firm willing to publish his sixty-page manuscript as a pamphlet.

Though a remarkable achievement, the paper had problems. In particular, Helmholtz struggled to fit the behavior of heat into the idea of energy conservation. Like William Thomson, Helmholtz saw merit in James Joule's experiments showing mechanical work and electrical energy can turn into heat. This suggested heat was simply another form of energy. But also like Thomson, Helmholtz was persuaded by Carnot's analysis of the steam engine, which was predicated on the idea that an unchanging quantity of heat generated work as it flowed from hot to cold. Heat, therefore, appeared asymmetrical. Other forms of energy could turn into heat, but heat, it seemed, could not transform into anything else. As Helmholtz put it:

"Whether by the development of mechanical energy, heat disappears, which would be a necessary postulate for the conservation of force, nobody has troubled himself to inquire."

This statement, though true, is rather unfair on Helmholtz's contempo raries. James Joule had measured the amount of work needed to generate a given amount of heat, but designing an experiment that went the other way and measured if heat "disappeared" as it did work was impossible in practice. It would involve measuring the heat that flowed out of a steam engine's furnace and then into its sink with a precision that was beyond 1850s technology.

The truth about the nature of heat and how it did work remained hidden.

# CHAPTER SIX

# The Flow of Heat
# and the End of Time

Over and above his direct contributions to science, Prof. Magnus exercised a powerful indirect influence, through the kindly aid and countenance which he lent to young inquirers.

—From the 1870 obituary in *Nature* of Gustav Magnus, written by the Anglo-Irish physicist John Tyndall

Students adored Gustav Magnus.

Unlike most Prussian academics, the Berlin University professor lectured in short sentences, "reminiscent of English." Magnus augmented his talks with striking physical demonstrations carried out on equipment he bought with money he'd inherited from his father, a wealthy merchant.

Magnus embodied the transformation in the way science was taught in the German-speaking lands in the first half of the nineteenth century. Universities here had pioneered the use of seminars. Unlike lectures, where professors held forth to large student audiences, at seminars small groups took part in free-flowing discussions with their teachers. To this end, Magnus invited ten or so elite scholars to a weekly "physical colloquium" at his home, a baroque mansion in Berlin's Mitte district. Attendees studied and then defended scientific topics from attacks by the others. The teacher engaged as an equal, never using seniority as a shield.

A few months after Helmholtz had published his tract on the conservation of energy, Magnus brought it before the colloquium. To scrutinize it, he selected Rudolf Clausius, a twenty-six-year-old from Köslin, a town in Prussia (now in Poland). The sixth son of a Lutheran pastor, Clausius had attended a school run by his father before going on to Berlin University, where he was awarded a doctorate for investigating the colors of the sky. Though the explanation in his dissertation was wrong, Clausius's gift for abstract reasoning swayed the examiners, a gift amply displayed in a career in which he eschewed experiments and instead sought truth with the aid of logic and mathematics. And although Magnus's graduates would dominate German science through the 1850s and 1860s, Clausius would outshine them all, becoming the father of theoretical physics.

No record survives of the colloquium at which Clausius spoke on energy conservation. But to prepare, he pored over the works of Helmholtz, Carnot, Thomson, and Joule and was finally able to solve the problem that had baffled his predecessors—how to reconcile the idea that heat might be a form of energy that can be converted into other forms, with Carnot's view that heat must flow from hot to cold to create work.

Clausius's solution? Both views are right. Heat can be both created and destroyed, *and* it must flow from hot to cold for work to be created.

Clausius described his breakthrough in a historic paper that appeared in *The Annals of Physics* in 1850.

With flawless logic, Clausius reasoned as follows:

Carnot had analogized heat engines to water mills. In the latter, water creates energy in the form of work as it streams downhill. In heat engines, caloric fluid was supposed to do the same as it flows from hot furnace to cool sink. The same amount of both substances leaves their respective devices as enters. Neither water nor caloric fluid are destroyed.

Clausius abandoned this comparison. Though water enables a mill, it does not turn into work. That derives from gravity. Water at a height possesses potential energy that becomes work with downhill flow. Clausius had learned this from Helmholtz.

In an engine, things are different. Thanks to Joule's work, Clausius was mindful that work can be turned into heat. But he then took a step that no one else had dared—he assumed the opposite was indeed true and that, yes, in an engine, some heat is turned into work. He then showed that this assumption did not contradict Sadi Carnot's ideas. They just had to be modified slightly.

Clausius reasoned as follows: Carnot had been wrong to say that *all* the heat flowing into an engine eventually flows out. But some does. This is not converted into work but is wasted. You can feel this if you put your hand near the exhaust pipe of your car. The warmth you sense is evidence that no matter how well engineered the system, some heat will always escape.

The reason for this is subtle. Imagine a simple engine consisting of a single cylinder in which expanding gas pushes a piston. In an internal combustion engine, heat is generated inside the cylinder by burning petrol or diesel. In Clausius's hypothetical engine, the heat flows in from an unspecified outside source and none is lost or wasted as friction. But no matter, this idealized engine helps illuminate the principles at work.

So, to start with, heat flows into the gas in the cylinder, causing it to expand and thereby push a piston. As this happens, per the law of energy conservation, heat *becomes* work. If the cylinder were infinitely long, the expansion would continue forever. In principle, all the heat could become work. But an infinitely long cylinder is an absurdity.

For an engine to keep going, some of the work created during expansion must be sacrificed, pressing the piston back to its starting position. To minimize that, the gas in the cylinder is cooled so it's easier to squeeze.

But as the piston returns, it squashes the gas in the cylinder. This reheats it, making it resistant again. For a sense of how this happens, squeeze a balloon full of air. You will feel it become hotter.

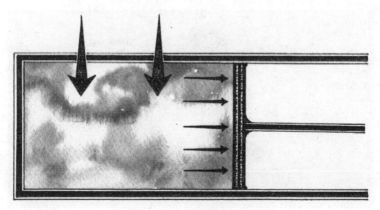

Heat makes the gas inside a cylinder expand so it pushes a piston.

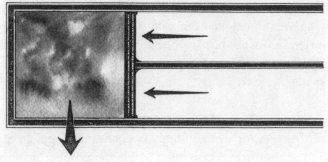

Heat is removed from the gas as the piston returns.

So, heat must be sucked out of the cylinder into a sink in this compressing stage. If this wasn't done, all the work created in the expanding stage would be used up. The engine would be useless. In a typical car engine this back-and-forth process is repeated rapidly. Heat is created in and then flows out of its cylinders several times every second.

Putting these principles together, this is how Clausius envisaged an ideal engine:

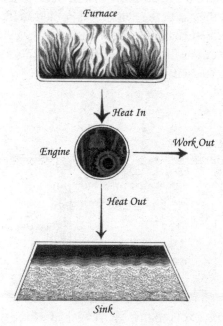

Clausius's ideal engine

In summary:

Initially as heat flows into the ideal engine from the furnace, all of it turns into work, per the principle of energy conservation. Then, for the engine to keep going, a proportion of that work is returned to the engine. This turns back into heat, also per the principle of energy conservation. This is the waste heat that's discarded.

To make the engine more efficient, make the furnace hotter. The gas in the cylinder then expands more forcefully and generates more work. Or you could make the sink cooler, which renders the gas more squeezable, meaning that less work is consumed during compression.

Conversely, the split worsens as the temperature difference between furnace and sink narrows. When it's zero, all the work created during expansion is sacrificed during compression. The engine does no work.

This was close to the pivotal conclusion Carnot had drawn from his engine/water-mill comparison—that the temperature drop between furnace and sink alone fixes the amount of work that an ideal engine derives from a given flow of heat. (For more details see Appendix II.)

But was this still always true? Might different substances divide heat into work and waste heat differently? Imagine an air-driven engine and a steam-driven machine, both operating between the same furnace and sink. Might the air-driven version create more work during expansion or sacrifice less work during compression than the steam-driven one?

Answering this required Clausius to discover a new law of physics.

Clausius started with a Carnot-inspired thought experiment. He imagined an ideal reverse engine. Work is done *to* such a device and it pumps heat *from* a cool place *to* a hot one, in other words from the sink to the furnace. This resembles a modern refrigerator, which shifts heat from its interior out into the room. But remember the law of energy conservation. The work done to the refrigerator must go somewhere and in fact it turns into heat—the opposite of an engine in which some of the heat flowing into it turns into work. Put your hand behind your refrigerator and the warmth you feel there is the sum of the heat leaving its interior *and* the heat generated by the pump.

Clausius imagined an ideal engine and refrigerator working between the same furnace and sink.

He stipulated that the work produced by the ideal engine drive the ideal refrigerator.

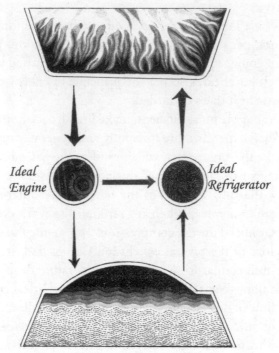

Clausius's ideal engine drives an ideal refrigerator.

For illustration, the ideal engine takes in one hundred calories from the furnace, turns half of that into work, and discards the other fifty calories into the sink.

The ideal refrigerator takes the fifty calories worth of work from the ideal engine, sucks fifty up from the sink, and pumps one hundred calories into the furnace.

This arrangement will run forever. All the heat leaving the furnace will be replenished. All the heat rejected into the sink will be lifted back out. But no net work will emerge.

Next, Clausius imagined a better-than-ideal engine. This divides the heat it receives into work and waste heat *more favorably* than the ideal engine. Instead of a 50:50 split, it operates on a 50:30 basis. Such a machine receives eighty calories from a furnace, turns fifty into work, and discards thirty into the sink.

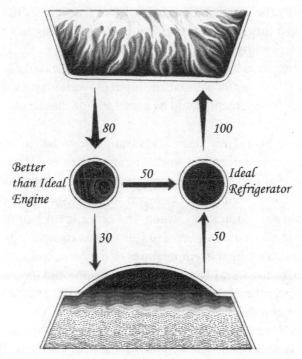

A better-than-ideal engine drives an ideal refrigerator.

Next imagine that the better-than-ideal engine drives the ideal refrigerator.

Eighty calories flow from the furnace into the better-than-ideal engine. Fifty of those calories are turned into work and thirty are dumped into the sink.

The ideal refrigerator uses the work from the better-than-ideal engine to suck fifty calories out of the sink. It then sends a total of one hundred calories back to the furnace.

This is the key. At no stage has the law of energy conservation been violated. The sum of heat and work has remained constant. But something strange has happened.

The furnace lost eighty calories of heat to the better-than-ideal engine but gained one hundred calories from the ideal "reverse" engine, meaning a net *increase* of twenty calories.

Meanwhile, the sink gained thirty calories from the better-than-ideal engine but lost fifty to the reverse ideal engine, resulting in a net *decrease* of (50 – 30) or twenty calories.

Overall, twenty calories of heat have flowed *from a cool place (the sink) to a hot one (the furnace)* without any input of work from outside the system. Such an arrangement would be a refrigerator that needs no power to operate.

This cannot exist. Heat never flows from cold *to* hot unless it's forced to—i.e., without some input of work. The spontaneous unforced direction of heat flow is always the other way around, from hot to cold. A better-than-ideal engine cannot exist because it would violate this principle.

Thus, Clausius vindicated Carnot. The Frenchman had been right to conclude that the maximum amount of work extractable from a given supply of heat is set by the temperatures of the furnace and the sink. It is independent of the working material of the engine and its design.

Carnot's hypothesis still stands if one accepts the principle of energy conservation *and* that heat never spontaneously flows from cold to hot.

The overriding conclusion of Clausius's paper was therefore that the behavior of heat is defined by two principles, now known as the first two laws of thermodynamics. These are:

First law: Though heat and work can be converted into each other at the fixed "exchange" rate that Joule had discovered, the total amount of heat plus work remains the same. (This is the conservation of energy as applied to heat and work.)

Second law: Heat never spontaneously flows from cold to hot.

These two statements marked the official birth of a field of science.

Clausius's paper was well received. After its publication, he was appointed professor of physics at the Royal Artillery and Engineering School in Berlin. An English translation of the paper appeared within a few weeks. In the summer of 1850, reading it in Glasgow, Thomson must have experienced mixed emotions. On the one hand, Clausius had acknowledged Thomson's role in bringing the ideas of Carnot and Joule to the attention of the scientific community. On the other hand, someone else had solved the puzzle Thomson had wrestled with for two years. Some months later, Thomson published his own derivation of Clausius's second law.

Although Clausius and Thomson never met, the pair in a sense worked together across the pages of scientific journals to establish deeper and deeper insights into the nature of heat, giving the nascent science a firm footing. Together they paved the way to showing thermodynamics had a broader relevance beyond steam engines.

The first and one of the most dramatic steps in this direction came in a paper Thomson published in April 1852. As had Joule, Thomson sensed the hand of the creator in the behavior of heat. In the first law of thermodynamics, the conservation of energy, Joule had noted "the grand agents of nature." In the second law, Thomson saw God's plan for the fate of the cosmos.

It's worth exploring the context in which Thomson did this work.

In Glasgow, despite its being an industrial boomtown, misery also abounded. The Irish potato famine had brought nearly one hundred thousand destitute refugees to the city. Hunger, primitive health care, and poor sanitation led to epidemics. No one was immune. In early 1847, Thomson's younger brother John contracted typhus in the infirmary where he was studying medicine and died within a week. Cholera followed typhus, and less than two years later Thomson's father succumbed, one of over four thousand deaths caused by the disease in the city that year. Soon after this second bereavement, Thomson was spurned by the young woman Sabrina Smith, to whom he had proposed.

Bereavement and rejection—these were the emotions Thomson was feeling when he wrote his paper of 1852, entitled "On a Universal Tendency in Nature to the Dissipation of Mechanical Energy." Clearly he was no longer thinking only about steam engines.

Four years earlier, in his 1848 paper on Carnot, Thomson had pondered what happens to an iron bar that's red-hot at one end and cool at the other. Heat flows out from the hot extremity until the bar's temperature is the same throughout. What happened, Thomson had wondered, to the work that this same heat flow could have produced if it had taken place in an engine?

Now, in 1852, Thomson had the answer. The initial temperature difference across the bar, instead of being turned into work, has become "dissipated" heat. It is no longer useful. Thomson stressed that in accordance with the law of energy conservation, no heat is destroyed; but by being distributed differently, by no longer being concentrated in one end of the bar, it has lost its potential to do work.

An iron bar equalizing in temperature can therefore be regarded as a heat engine whose efficiency is zero. In an ideal machine, a proportion of the heat is turned into work and the other proportion is discarded. In the iron bar, the fate of all the heat is to be discarded. In both cases, once lost, it can no longer be used to create work.

If nature abhors a vacuum, heat abhors variation. Heat will always tend to dissipate and even out temperature differentials. Although it is possible to obtain some work as this takes place, any such gain, Thomson argued, would be temporary. The destiny of the work created is also to end up as dissipated heat. This takes place through unavoidable frictional forces such as the rubbing of wheels on a road surface or the drag of the water on a ship's hull. Worse, this process cannot be undone. Nature allows the transformation of work into dissipated heat. It does not allow the reverse.

In this paper, Thomson homed in on an idea that was new to physics in the mid-nineteenth century—irreversibility. This concept was absent from Newton's work—his laws are reversible. Say, a person drops a ball of a known weight out of a window at a known height. By applying Newton's laws, a second person on the ground can calculate the speed at which the ball was moving downward the instant before it hits the ground. That second person then throws the ball back up at that speed. The ball will return to the thrower. When a wheel rubs along a road, however, it's very different. Some of the mechanical energy of the wheel is turned into heat by friction. It's impossible, argued Thomson, to retrieve that frictional heat and give it back to the wheel.

In the science of heat, Thomson saw the same irreversibility he had witnessed in his life. There is a reason so much of what happens in the universe goes in one direction and not the other. It is a manifestation of the one-way dispersal of energy, which also lies behind the flow of time from past to future. Thomson had discovered time's arrow: it pointed irreversibly from a less dissipated past to a more dissipated future . . . and once that dissipation is complete, time ends.

Thomson had turned a cooling iron bar into a metaphor for the universe. All change in the cosmos was due to pockets of concentrated heat dissipating. In an early draft of his paper, Thomson writes, "I believe that no physical action can ever restore the heat emitted from the sun, and this source is not inexhaustible. . . . 'The earth shall wax old etc.' The permanence of the present forms and circumstances of the physical world is limited."

Thomson had synthesized his theology, his experience, and his science. In the final publication, he removed the biblical reference but was no less bleak about the future: "Within a finite period of time past, the earth must have been, and within a finite period of time to come, the earth must again be, unfit for the habitation of man as at present constituted."

This idea that the universe will wind down and die as all the heat in it dissipates became known as the heat death of the universe. Thomson's contemporaries noted the audacity with which he had drawn a grand cosmological conclusion from his everyday observations of a cooling iron bar. As Hermann Helmholtz wrote in 1854, "We must admire the sagacity of Thomson, who in the letters of little known mathematical formulae, which speaks only of the heat, volume, and pressure of bodies, was able to discern consequences which threatened the universe, though certainly after an infinite period of time, with eternal death."

# CHAPTER SEVEN

# Entropy

Die Entropie der Welt strebt einem Maximum zu.

—Rudolf Clausius

Predicting the end of time wasn't enough for William Thomson. Soon after, he conceived an idea that wrote his name into the scientific lexicon. This was the so-called absolute temperature scale.

Then, and to a large extent now, the way substances like mercury expand when heated serves as a measure of temperature. But this can lead to pitfalls as the following example shows:

Put a mercury based thermometer in a fridge. It reads 1°C. Take it out of the fridge. The mercury expands, moving up the thermometer's column till it reads 4°C. (In this example, although the kitchen air is warmer than the fridge's interior, it's a cold day.)

These readings are based on the convention that when mercury expands by around 0.018 percent, it signifies an increase of one degree Celsius. (The thermometer column is very thin to ensure that such a small volume change is perceptible.)

But how reliable is this method of gauging temperature? Would a different substance show the same change in temperature? For example, imagine a thermometer with water in its column. Perform the same measurements as you did earlier.

Now you find that instead of expanding, the water shrinks when taken out of the fridge. The actual temperatures of the fridge and the kitchen haven't changed, but the behavior of the substance used to measure them is very different. The mercury says the room is warmer than the fridge's interior, while the water appears to suggest the opposite. Which substance should you trust?

*"Puffometer"*

Using a water mill to measure height

Thomson saw how to free temperature from a material's propensity to expand or contract as it's heated or cooled. In other words, he saw how to devise an "absolute" temperature scale. It required he think of an ideal Carnot engine as a thermometer. To follow his logic requires some mental yoga.

Picture a tower in the square of a medieval village. The inhabitants wish to install windows at equal distances down its height. But they have no reliable yardsticks. However, they do possess a portable water mill and are able to measure the amount of work it generates in units called PUFFs.

The village's engineers install a water tank at the top of the tower. Just below, they position the mill. They let water flow through it, thus producing work.

In tiny steps, workmen lower the mill until it does one PUFF of work. They mark that point on the tower and define it to be a distance of "one degree" below the top.

The mill slides down farther until the "puffometer" reads two. That point is two degrees lower than the summit.

And so on. For every extra PUFF of work created, the drop from the apex of the tower is declared to be one degree greater. The villagers install the tower's windows five degrees apart, confident they are equidistant.

Thomson applied this methodology to defining temperature, substituting an ideal heat engine for the water mill and degrees of temperature for those of height.

To start with, the furnace and sink of the engine are at the same temperature. No heat flows. The engine does no work.

Lower the sink's temperature till the engine creates one PUFF of work. Define the sink to be one degree colder than the furnace.

Keep going. When the engine creates two PUFFs of work, the sink is two degrees colder. Three PUFFs, three degrees, and so on.

This engine can operate as a thermometer.

To measure, say, the temperature of a freezer, use it as the engine's sink. Note the quantity of work produced. If it's one hundred PUFFs, you know the freezer's temperature is one hundred degrees below that of the furnace.

This is an absolute value that's independent of any material's thermal properties.

Conceptualizing temperature in this way brings little practical benefit. Ideal engines are unbuildable, and the notion that one needs to fire up an engine when assessing temperature is preposterous.

But this is an advance. Remember, in an engine, some heat flowing in from the furnace becomes work, some is discarded. As the sink's temperature falls, this split becomes ever more favorable.

Then there comes a point when the sink is so cold that all the heat from the furnace becomes work.

This is the end of the line. An engine can do no better than turn all the heat it receives into work. Otherwise it would be generating work from nothing in violation of the law of energy conservation. The temperature of the sink at which this occurs is therefore the *lowest possible temperature that can exist.*

Our universe has limits. The speed of light, which nothing can exceed, is one example. Thomson's lowest possible temperature is another.

This is "absolute zero," and it shed light on a phenomenon that many scientists had observed but had struggled to explain—the effect of temperature on how much volume a gas occupies. Fill a balloon with air and then cool it at sea level so the air pressure on it doesn't change. As the temperature falls, the air shrinks. Also, the rate at which the air contracts goes up the colder it becomes. So a 50°C drop from 50 to 0 will shrink a volume of air by a greater amount than a drop from 100 to 50.

Nineteenth-century scientists could cool gases down to around −130°C but were unsure what happened beyond that. Some liquefied. Others, notably oxygen and nitrogen, did not, continuing to shrink. Extrapolating the graph of the relationship between temperature and gas volume suggested that at a temperature of −273°C, a gas would occupy no space and also, therefore, exert no pressure.

This chimed with Thomson's deduction that in theory there is a temperature at which an engine wastes no heat.

If the gas in the cylinder of an ideal engine is at −273, it exerts no resisting pressure. It will take no effort to push the piston back to its starting position.

Based on this logic, it was reasonable for Thomson to surmise that the zero of his absolute scale was −273, the same as the one suggested by gas behavior. For the sake of convenience, he defined the size of a degree in his scale to be the same size as a degree Celsius.

A century later, at the tenth General Conference of Weights and Measures in the town of Sèvres near Paris in 1954, the participants declared that the absolute scale should be named in Thomson's honor. As by then he was known by his title, Lord Kelvin, its units are called kelvins. The latest measurements deem that in this scale, −273.15°C is 0 kelvin, and that at sea level, the melting point of ice is 273.15 kelvin, and the boiling point of water, 373.15 kelvin.

Thanks to Thomson, temperature can be seen as a fundamental property of any object, just like its mass. Different objects, whether a fried egg or a nugget of gold or a volume of air, all weigh a certain number of kilograms, irrespective of what they consist of. The Kelvin scale permits a similar claim for their temperature. Just as with mass, physicists can investigate the behavior and effects of temperature with mathematical equations confident that its definition is not contingent on the arbitrary properties of a substance. We can even speak of black holes having temperature.

Through the 1850s, Clausius remained hard at work in Berlin and then in Zurich, bringing insight after insight to the way heat dissipates. The fruit of this labor was the definition of a new concept, *entropy*, a physical quantity that matches energy for importance. It was a secret buried in the way heat flows.

Imagine a large house with many rooms. Some possess radiators and are warm. Others lack heating and are cold. All the walls are insulated, and adjoining doors are shut.

Turn off the radiators and fling open the doors that connect the rooms. Heat flows out of the warm rooms and into the cold ones. Soon every part of the house will achieve the same temperature.

Clausius conceived of entropy to capture mathematically this way in which heat strives to redistribute itself. In the house example, he, in effect, declared it to be a measure of how spread out the heat is within its walls. At first, most of it is concentrated in a small number of rooms. Many others are cold. The heat is "undispersed." There is a great deal of temperature variation. Clausius defined the entropy of such an arrangement to be low.

After opening the doors, the heat spreads and the temperatures of the rooms begin to equalize. Following Clausius's definition, the entropy of the house rises. The less the variation in temperature and the more evenly distributed the heat, the greater the value of the entropy.

To grasp how entropy changes with the flow of heat, picture a house with only two rooms, one hot one and the other cold.

Entropy is a measure of heat dispersion. That means each room has its own entropy, representing the amount of heat dispersed within it. Call these *Entropy(hot room)* and *Entropy(cold room)*.

The entropy of the whole two-room house is *Entropy(hot room) + Entropy(cold room)*.

The door opens. Heat flows. The hot room becomes colder, the cold one warmer.

Less heat is now dispersed through the hot room. I.e., *Entropy(hot room)* has fallen.

But more heat is dispersed in the cold room. I.e., *Entropy(cold room)* has risen.

Clausius defined *changes* in entropy as follows:

When a quantity of heat flows out of a hot room, it will always cause that room's entropy to fall by a smaller amount than the increase in entropy caused when that same amount of heat flows into a colder room.

In the two-room example, therefore, as the heat flows, the *increase* in *Entropy(cold room) will be greater* than the *decrease* in *Entropy(hot room)*.

And that means, the entropy of the whole two-room house has gone up.

By defining entropy in this way, Clausius found a mathematical way of stating his law that heat always flows from hot to cold unless prevented from so doing. In any system that's insulated from the outside world, entropy will always tend to increase.

Algebraically, this is written as $\Delta S >= 0$. This short equation is one of the most important in all science. $\Delta$ is the Greek letter delta, which often symbolizes change in mathematics; $>=$ means "greater than or equal to." $S$ is the letter Clausius chose to represent entropy. There's a charming, unsubstantiated story that he did so in honor of Sadi Carnot.

The idea that the same amount of heat causes a greater change of entropy in a cold place than a hot one can seem strange. But consider this as an analogy: A noisy, crowded pub is next to a quiet library. Five rowdy people leave the pub. The din drops but by an indiscernible amount. The five stumble into the library. The noisiness there increases noticeably. When a group of raucous people enter a quiet place, the increase in disruption there is much greater than its fall in the boisterous place from which they came.

Similarly, when a quantity of heat flows out of a hot room, the fall in entropy there is smaller than the increase that occurs when that heat enters a cold room.

In summary, to say the entropy of a system increases is to say the heat within it is becoming more widely dispersed.

But, though Clausius's equation says this always tends to happen, it doesn't specify the rate at which it does so.

Insulating the walls of the rooms and keeping the doors shut can slow down almost to a standstill the rate at which entropy goes up.

Thinking in this way has another benefit. It helps us understand engines as devices that exploit low entropy.

Replace the open doors in the house with heat engines. Heat flows through them as it moves from hotter to colder rooms. In each engine, some heat is turned into work—perhaps it pumps water out of a mine. The rest disperses. Eventually, the rooms' temperatures equalize. Once the house reaches this state of maximum entropy, the engines will stop working. The heat in the house will no longer be of any use.

Increasing entropy is thus a measure of the decreasing usefulness of heat.

All this can seem fanciful. But the multichambered house is a way of

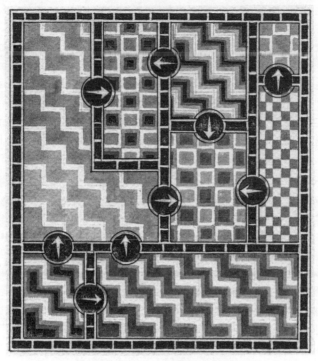

Using engines to exploit low entropy

understanding any system in which heat disperses. It is a simulacrum for the modern world. We release concentrated heat from fossil fuels, atomic nuclei, sunshine, geothermal sources, or wind. As it flows, we turn some into work that enables our homes, factories, and transport.

Life, too, runs on this principle. Plants live by dispersing solar energy, animals by dissipating calories from food.

$\Delta S >= 0$ rules us all.

In 1865, Clausius revisited the two laws of thermodynamics that he had first stated in his paper of fifteen years prior. He updated them by employing the word *energy* instead of *Kraft*, and he added his own coinage, *entropy*. The laws state:

1. The energy of the universe is constant.
2. The entropy of the universe tends to a maximum.

(*Universe* means any system that's closed or sealed off. But because the universe we live in has nothing beyond it, it is true that its energy cannot change and its entropy tends to rise. More intuitively, the second law can be stated: the entropy of any closed system tends to increase.)

These two lapidary statements are a testament to the human intellect and imagination. They are a scientific milestone every bit as significant as Newton's laws of motion, which were published two centuries earlier.

Since 1865, when Clausius published these principles, they have been at the forefront of physics, helping humans better understand everything from atoms to living cells to black holes. They are a remarkable testament to the power of the human intellect and imagination.

But scientific principles can be misunderstood and misused even when they are true. At the moment thermodynamics arrived on the world stage, it found itself turned into a weapon to fight the latest ideas in another field of science. The protagonists were two giants of nineteenth-century British thought—William Thomson and Charles Darwin.

"One species does change into another." Charles Darwin scribbled this phrase into a notebook in 1837 after his five-year voyage as a naturalist on board the HMS *Beagle*. What he'd seen in South America, the Galápagos Islands, and elsewhere had persuaded him that species mutated into one another, and that competition for resources and other environmental pressures drive this process. Aware of the enormity of his ideas, Darwin spent the next two decades collating data to support them. As well as the specimens he had observed and collected himself on his travels, he scrutinized the works of other naturalists, studied domestic animal breeding at length, and documented variations in the forms of barnacles, beetles, and finches.

One assumption that this most careful of scientists felt was unassailable was that the earth was immeasurably old. Darwin had with him on the *Beagle* volume 1 of *Principles of Geology*, by the Scotsman Charles Lyell, who is still considered one of the founders of this branch of science. The book espoused the idea of uniformitarianism, which stressed that natural processes such as weather, tides, and erosion by wind and water have gradually shaped the surface of our planet. Moreover, argued Lyell, as these processes are incredibly slow—consider how a coastline erodes, for example—the earth must be many hundreds of millions of years old. This idea stood in opposition to catastrophism, which claimed that the earth was young and shaped by sudden planetwide violent events— earthquakes, eruptions, and floods of much greater power than humans had ever observed. In Britain particularly, many writers found this latter view chimed with biblical descriptions of events such as Noah's flood.

Darwin understood that evolution through natural selection needed eons, and Lyell's uniformitarian arguments provided them. In *The Origin of Species*, published in 1859, Darwin accepts that if the earth is young, his theory fails. Anyone who disagrees with uniformitarianism and its doctrine of an ancient, slowly changing planet "may at once close this volume." When writing these words, he knew theologians or geologists still holding catastrophic views would find fault with his ideas. But he never suspected that the mathematical equations of thermodynamics would torment him till his death twenty-three years later.

Since his undergraduate years, William Thomson had had in interest in geology and whether physics could illuminate its mysteries. In his time, the distinctions between such disciplines hadn't formally emerged, and he believed physical principles discovered in a laboratory applied to the earth and the universe as a whole. By the 1850s, he had become convinced the doctrine of uniformitarianism was flawed, and he published his objections in scientific journals, which few members of the general public read. Then in late 1860, a year after the publication of *The Origin of Species*, Thomson broke his leg. Bedridden and with time to think, he concluded that if uniformitarianism was in error, so, too, was Darwin. He sent off the first of many critiques of the English naturalist's theory to *Macmillan's Magazine*, a publication aimed at bringing contemporary literature to the educated layman. (Its contributors included Alfred Tennyson and Rudyard Kipling.)

Thomson was a man of faith but he had no truck with biblical literalists who believed the earth to be six thousand years old. His position was that a slowly changing ancient earth stood in direct contradiction with the scientific principles that he had worked so hard to establish—that energy cannot be created or destroyed and that heat tends to dissipate. Using these laws, argued Thomson, it would be possible to estimate the age of the earth and investigate whether it was old enough for evolution to take place.

In April of 1862, he brought out a paper claiming that a thermodynamic analysis of the flow of heat in the earth showed directly that it must be younger than uniformitarians, and by extension Darwin, believed. It starts, "Essential principles of thermodynamics have been overlooked by geologists." Dissipation was the key to Thomson's argument. Observations from mine shafts and tunnels showed that the earth's temperature increases with depth below the surface. Thomson's friend the Scottish physicist J. D. Forbes, by taking measurements in and around Edinburgh, estimated that the earth's temperature rose by 1°F for every fifty feet of descent. This persuaded Thomson that the earth was cooling, losing heat to the atmosphere.

Using elegant mathematics, Thomson combined Forbes's measurements with others relating to the thermal conductivity and the melting point of rock. Even acknowledging uncertainties in the data, he concluded the earth's age was somewhere between 20 million and 400 million years. This was far too short a time for evolution. Even if the older estimate was true, Thomson argued the earth would have been considerably hotter than it is now for most of its existence. Before around 20 million years ago, the temperature of the entire earth would have been so high that the whole globe was molten rock. Evolution's requirement that the earth was much as it is now for eons defied thermodynamic sense.

Darwin was shaken. "Thomson's views on the recent age of the world have been for some time one of my sorest troubles." "I am greatly troubled at the short duration of the world according to Sir W Thomson." "Then comes Sir W Thomson like an odious spectre"—these are lines from Darwin's letters to friends. In turn, his allies felt unqualified to attack the physicist's arguments and suggested that perhaps evolution worked faster than previously believed, a solution that didn't satisfy Darwin.

In his lifetime, Darwin never found a satisfactory response to Thomson. His final published statement on the matter accepts that Thomson's objection to the theory is "one of the gravest as yet advanced." He goes on, "I can only say, firstly, that we do not know at what rate species change, as measured by years, and secondly, that many philosophers are not as yet willing to admit that we know enough of the constitution of the universe and of the interior of our globe to speculate with safety on its past duration."

Though Darwin didn't live to see his hunch about earth's antiquity vindicated, Thomson did learn of his mistakes. In the 1890s, a new source of energy, radioactivity, was discovered. This led to an understanding that vast amounts of energy are hidden within atoms. Although it would be some decades before the mechanism of stellar nuclear fusion was understood, measurements of energy released by radioactive elements showed that a body as large as the sun could emit heat for billions of years. Furthermore, the new technique of radioactive dating revealed that the earth possessed rocks that must be of a similar age. One of the pioneers of the new radioactive science, Ernest Rutherford, gave a speech in 1904 at the Royal Institution in London, in which he declared, "The discovery of the radioactive elements . . . allows the time claimed by the geologist and biologist for the process of evolution."

Thomson, now aged seventy-nine, was unconvinced and remained so until his death three years later.

This is how science works. Thomson was no more or less stubborn than anyone else. Scientists are human and are motivated by emotional attachments to ideas and guided by intuition as all creative people are. These can be powerful tools that lead to the truth, but they can also mislead. Thomson felt in his bones that Carnot was right about the importance of the flow of heat, and he sought evidence to justify that hunch. In the early years the evidence was scant, but it was enough to enable Thomson to put down the foundations of thermodynamics. The same scientific instincts, the same intuition, told him Darwin was wrong about the age of the earth, and Thomson sought evidence to support this view. In the case of Carnot, Thomson was right; in the case of Darwin, he was wrong.

It's a mark of the stature accorded to Thomson and physics in the 1860s that his views were taken to heart by scientists in very different fields. The laws of thermodynamics were a triumphant acclamation of the industrial age. Heat was driving the world into the future, and scientists knew how heat behaved.

But what is heat? That question remained unresolved.

# CHAPTER EIGHT

# The Motion We Call Heat

How then does it happen that tobacco smoke in rooms remains
so long extended in immovable layers?
—Dutch meteorologist Christoph Buys Ballot

Rudolf Clausius's observation that heat spontaneously flows from hot to
cold led to the most accurate description yet of how heat behaves. But in
his work, he was at pains to point out that he wasn't relying on any specific theory as to what heat might be. Though it's clear from his later writings that he privately held views on the nature of heat, he feared that if he
expressed them publicly and they turned out to be wrong, it would diminish all that he'd achieved.

In 1857, something changed. An alternative explanation to the caloric
theory of heat, which came to be known as kinetic theory, had bubbled up
to the surface. Enough people were writing about it to prompt Clausius,
as the preeminent thinker on heat in mid-nineteenth-century Europe, to
write his own paper stating what he thought heat might be.

To appreciate Clausius's approach, think of the difference in temperature between summer and winter. And as Clausius had moved from Berlin
to become professor of mathematical physics at Zurich's brand-new university, the Eidgenössische Technische Hochschule or the ETH, imagine a
summer's day in that city. Records reveal that the temperature in July during daylight hours hovered around 20°C. By December, as the sun's rays
grew feebler, the temperature at midday barely reached a chilly 0°C. What
is different about air at different temperatures? Air *looks* the same in July
and December but *feels* markedly different. Why?

• • •

The basic tenets of Clausius's answer had been floating around at the periphery of scientific enquiry since 1738, when it was mooted by a Swiss polymath called Daniel Bernoulli. For the subsequent century, it had languished largely unnoticed in the shadow of caloric theory. In many ways Bernoulli's idea was too advanced for its time, and even he considered it something of an afterthought, probably because his main interest was in blood, rather than heat.

Bernoulli lived in an age when academics served at the pleasure of monarchs, and at twenty-five, he had been invited by the empress of Russia, Catherine the First, to be professor of mathematics at St. Petersburg University. There Bernoulli, who had also studied medicine, became fascinated by the way blood flows around the body. He noticed that if he inserted one end of a thin glass tube into an artery of his patient's forearm, the blood rose in the tube to a height of several inches. By measuring this height, Bernoulli deduced the patient's blood pressure. (This remained the common method for measuring blood pressure until the 1890s.)

But Bernoulli was a polymath—a doctor, mathematician, and physicist—and he decided to investigate further with a series of experiments. He simulated what was happening in his patients' arms by pumping water down a narrow pipe in which he had punctured a small hole. He then inserted a straw into the hole. Just as with blood flow in a human arm, the water rose to a certain height in the straw, indicating the pressure exerted by the water flowing down the pipe. But what surprised Bernoulli was that the height the water reached in the straw fell as the speed of the flow of the water in the main pipe increased. The *faster* the flow of the water in the pipe, the *lower* the pressure it exerted. Intrigued, Bernoulli turned to mathematics for an explanation.

Specifically, Bernoulli looked first at the one aspect of the physical world the mathematics of the day could describe, namely the way solid objects such as cannonballs and billiard balls move—the so-called principles of mechanics. These had been established by Isaac Newton in the 1680s and had become a sophisticated tool by the time Bernoulli was writing in the 1730s.

Bernoulli applied these principles to fluids such as blood and water and showed that they predicted what he'd observed. Now known as Bernoulli's principle, the idea can be extended beyond liquids to gases. You can see it in action every time you fly in an airplane. Look carefully at the wing and you will see its upper surface is curved and its lower one is flat. This shape forces the air flowing below the wing to move more slowly than the air above it, so that the upward pressure on the wing exceeds the downward pressure and the plane feels an upward "lift."

In 1738 Bernoulli published a book named *Hydrodynamica*, in which he explored the various implications of Newtonian mechanics to fluids. In chapter 10, entitled "On the Properties and Motions of Elastic Fluids, Especially Air," he tackled the question of what happens to a gas as its temperature changes. This, too, can be explained, he argued, with the aid of Newtonian mechanics.

First, Bernoulli considers the way gases resist as one tries to compress them. If you try to squeeze a balloon, for example, it takes some effort as the air inside pushes back; that is, the air exerts pressure. To explain this, Bernoulli conjectured that a gas consists of "very small particles in very rapid motion." He argues that though these particles are too small to see, they act like tiny billiard balls. So in the balloon example, the pressure you feel as you try to squeeze it is caused by these tiny particles colliding into the inner walls of the balloon. As each particle hits the balloon wall, it pushes the wall outward by a tiny amount. Although the impact of one air particle is imperceptible, the collective effect of all the particles hitting the wall is what we perceive as air pressure. Because the air particles follow the same principles of Newtonian mechanics as billiard balls, Bernoulli was able to derive a precise mathematical relationship between the pressure the air exerts and the volume it occupies if the temperature remains unchanged. Staying with our balloon, he predicts that if you squeeze it down to half its original volume, the pressure the air inside exerts will double. If you squeeze it to a third, it will treble; a quarter, it will quadruple; and so on. And this prediction is indeed borne out by experiment.

But what if the temperature does change? Bernoulli noted that heating a gas increases the pressure it exerts. With the balloon, if you heat it, the air inside will expand as it presses against its rubber walls with greater force. And if you take an inflated balloon and put it in a fridge to cool it down, you will see it shrink in size as the pressure exerted by the air inside it drops.

Bernoulli reasoned as follows: if heating a gas causes it to exert greater pressure, and if pressure is a direct result of the speed with which the air particles move, then adding heat to a gas must cause these particles to zip about at greater speeds. In other words, hot air feels hotter than cool air because particles in hot air zip about at a much greater velocity than in cold air.

This is a major insight, one that feels in hindsight like a landmark moment. When these "high-speed" particles hit our skin, we interpret that feeling as "hot." When sluggishly moving particles collide with our skin, that's "cold." Temperature is therefore a manifestation of the speeds at which gas particles move. The faster they move, the greater the temperature. And so, the difference between summer and winter air is down to the comparative speeds at which the air particles move. Or in Bernoulli's words, on a hot day they "are in more violent motion" than on a cold day.

Hindsight, however, is no guide to how Bernoulli's contemporaries responded to "the kinetic theory of gases." The truth is eighteenth-century physicists paid it little heed. Perhaps without any urgent need for scientific inquiry—improving steam engines, say—they had no need to do so. It is a prime example of how a scientific theory, however good it is, may disappear from view if it has no cultural, social, or economic relevance to society. Bernoulli's writings on heat were written over a century too early.

Still, Bernoulli's kinetic theory survived, cropping up now and again on the fringes of science over the next century. And by the mid-1800s, as the steam engine focused more and more minds on the nature of heat and caloric theory fell by the wayside, kinetic theory came back into view. Versions of the theory appeared in journals published in Manchester and Berlin, and in 1845, an English schoolmaster in Bombay submitted his take on kinetic theory for publication in the journal of the Royal Society, only to have it flatly rejected. "The paper is nothing but nonsense," wrote the society's referee.

Attitudes were soon to change, for in Zurich in 1857, Rudolph Clausius published his own paper, "The Nature of the Motion We Call Heat," supporting kinetic theory.

Kinetic theory appealed to Clausius's approach to physics. As a theoretician and a gifted mathematician, he felt that progress in physics happens as much in the mind as in the laboratory. Not only did kinetic theory allow him to envision the physical world at a minute and invisible scale, but it also enabled him to make a wide range of predictions about how all gases behave.

The style of Clausius's paper is remarkable. Only twenty-seven pages long, its first two-thirds contains no mathematics at all. He persuades the reader in ordinary language before backing up his reasoning with formulae and algebra in the paper's last third.

First, Clausius restates Bernoulli's hypothesis that the temperature of a gas is the outward manifestation of how fast its constituent particles are moving. But then he adds a subtle caveat, arguing that the particles are not always simple, featureless spheres. They can clump together to form complicated structures. For instance, they can form tiny dumbbells with a single atom at each end. This structure can also oscillate, with each end of the dumbbell moving closer and farther from the other end. This reasoning leads Clausius to a distinction between the temperature of a gas and the total amount of heat energy it contains. All the different kinds of motion added together represent the heat energy of the gas. But only the straight-line movements of the particles, which Clausius calls their "translatory motion," contribute to its temperature.

This explains a commonly observed aspect of the world we live in—namely that the same amount of heat raises the temperatures of different substances by different amounts. For instance, let the heat released from burning a kilogram of coal flow into two boxes, one containing molecules of helium gas and the other containing an equal number of oxygen molecules. The heat will raise the temperature of the helium by a greater amount than the oxygen.

We now know what Clausius only suspected: the structure of the respective molecules does indeed determine the types of motion they are capable of. Helium molecules are featureless spheres. Oxygen molecules are like the tiny dumbbells described earlier.

That means that when the box full of helium is heated, all the energy goes to increasing the particles' straight-line movements. In the oxygen box, the energy is partly used up making the particles oscillate more quickly, leaving less energy available to speed up their straight-line movements. Thus, the same amount of heat makes the helium molecules move faster than the oxygen ones and so raises the helium's temperature by a greater amount.

Clausius also extended Bernoulli's theory about the particulate nature of gases to liquids and solids, reasoning that all matter consists of trillions of particles in constant motion. In solids, these molecules vibrate around a fixed position. In liquids, Clausius theorized, the particles are in a constant flux, making bonds and breaking them at the same rate to produce the fluid form. In a gas, the molecules are completely free to move independently and in any direction.

Armed with these descriptions, Clausius then paints a compelling picture of how evaporation occurs. We've all seen that if we leave a bowl of water out for a few hours, even on a mild day, a noticeable amount of it disappears into the air. The reason is what's happening on the surface of a liquid. Here the bond making and breaking happens as it does throughout a liquid, but there are no particles to bond with above the surface. So every now and then, a particle on the surface breaks free from a particle below. If it is moving upward, it will find nothing to bond with and break free of the liquid. Over time, more and more particles inevitably follow suit and what was a full bowl of water has largely disappeared by evaporation.

Clausius's grasp of this relationship between the movement of gas particles and temperature meant that he could even predict the average speed at which molecules of the common gases in the earth's atmosphere such as oxygen and nitrogen move. In the case of oxygen at 0°C, Clausius estimated the average speed of each molecule as 461 meters per second—that is, well over a thousand miles an hour. The figure is within 1 percent of modern estimates.

Unwittingly, Clausius had also come up with an explanation for why the atmosphere is able to retain large quantities of the gases oxygen and nitrogen. Newton's laws allow us to calculate what's known as the earth's escape velocity. Anything that travels faster than this speed, which is around 11 km/sec, will escape the earth's gravity and be lost into space. As the average speed of an oxygen molecule is a tiny fraction of the earth's escape velocity, at only 0.5 km/sec as opposed to 11 km/sec, it remains in the atmosphere. On smaller celestial objects, where the escape velocity is closer to or less than the average speed at which gas molecules move about, they drift off into space, leaving the object bereft of an atmosphere. Calculations such as these help modern-day astronomers seek out planets that might be habitable.

Clausius's 1857 paper is a wonderful example of how great science takes us under the surface of the world we live in—it proposes explanations for a cool breeze and why there's oxygen in our atmosphere to breathe.

Ironic, then, that kinetic theory was nearly derailed using similar think-ing. For in 1858, just a year after Clausius published, a Dutch meteorolo-gist named Christoph Buys Ballot wrote a paper arguing that a common observation about gases in the real world contradicts Clausius's claim that their constituent molecules fly about at high speeds. Buys Ballot argued that, if gases behaved as Clausius proposed, one would expect them to mix rapidly. So if a bottle of a strongly odorous gas such as chlorine is opened in one corner of a room, its molecules should reach the other side of the room in around one one-hundredth of a second. One would expect to smell the chlorine instantly on the other side of the room instead of hav-ing to wait several minutes, as happens in reality. Buys Ballot summed up his objection to Clausius's theory with a vivid observation—"How then does it happen that tobacco smoke in rooms remains so long extended in immovable layers?" You can almost hear him saying, as he lights another cigar, "Aha, gotcha now!"

Clausius was a good enough scientist to take seriously a criticism that made sense. His response to the "tobacco smoke" problem was a second paper on kinetic theory with a clever twist. He argued that the size of the molecules, though small, was significant. Gas molecules are basically tiny spheres or clumps of spheres in constant collision, like frantic bum-per cars. Although an individual molecule is moving rapidly, it has barely gone any distance before it collides into another one, and its journey is deflected. It collides again afterward, which causes another deflection, and so on. Molecules don't therefore travel in straight lines. Instead they zig-zag, going backward, forward, and sideways. Although their speed from collision to collision is high, it still takes a long while for molecules to cover any significant distance.

Clausius's view of gas behavior is taught in high school physics classes today. But because we grow up with this description, it's hard for us to comprehend the mindset of those contemporaries who found Clausius's papers unsatisfactory. The problem was that they made all sorts of pre-dictions that couldn't be verified. Gases were made of molecules in con-stant motion, Clausius claimed, but he showed no way to estimate their size. He made predictions about the average speed of molecules, but these were also unverifiable. And without new and testable predictions, kinetic theory remained an unproven hypothesis.

Fortunately, however, Clausius's papers were translated into English.

And in February of 1859, the English version of Clausius's second paper on kinetic theory landed on the desk of a certain physics teacher at Marischal College in Aberdeen in the far northeast of Scotland. That happenstance proves that, sometimes, as important as who writes the words is who reads them.

# CHAPTER NINE

# Collisions

The true logic of this world is the Calculus of Probabilities.
—James Clerk Maxwell

In February, the average temperature of the water in the North Sea off the town of Aberdeen on the northeast coast of Scotland is 6°C. The water temperature anywhere off the British coast at that time of year is uninviting, but the North Sea, largely untouched by the warming waters of the Gulf Stream, is particularly so. Aberdeen also lies at a northerly latitude of fifty-seven degrees, which means that even at midday the winter sun barely reaches a height of ten degrees above the horizon and is only visible for eight hours. The air above the water hovers around freezing.

In February 1857, at much the same time that Rudolf Clausius was pondering the physics of why things feel hot or cold, the newly appointed professor of natural philosophy at Marischal College in Aberdeen, a twenty-five-year-old named James Clerk Maxwell, could be seen strolling at the foot of the black cliffs that line the coast just south of the city. Here, he stopped, removed some of his clothes, and plunged into the icy waters, taking his "second dip of the season." Invigorated, he followed this "with gymnastics on a pole afterwards."

James Clerk Maxwell was the youngest professor by fifteen years at the venerable Marischal College, which by the 1850s was housed in a vast baronial-style granite building that still dominates the city of Aberdeen. For Maxwell to obtain his professorship, he'd needed approval from the Lord Advocate and the Home Secretary of the British government. They had given it readily as his talent for physics and mathematics had been clear for some time. Born in Edinburgh in 1831 to a moderately wealthy landowning family who counted members of the eighteenth-century Scottish enlightenment as their forebears, Maxwell had written his first scientific paper, a treatise on Cartesian ovals and ellipses, when he was only fourteen. It was accepted by the Royal Society of Edinburgh, but because of his age, Maxwell was barred from attending the meeting at which Scotland's leading scientists had gathered to discuss it. He would have found the event overwhelming—he'd grown up a solitary and awkward child as his mother had died of abdominal cancer when he was eight and his early education had come from private tutors at his father's estate, Glenlair, in the southwest of Scotland. His boyhood friend Lewis Campbell remembered him as a nearsighted young man who, though earnest and well-meaning, found social discourse hard going—"His replies in ordinary conversation were indirect and enigmatical, often uttered with hesitation," Campbell recalled, and "when at table he often seemed abstracted from what was going on, being absorbed in observing the effects of refracted light in the finger-glasses."

Yet by the time he arrived to teach at Aberdeen, after studying at the universities of Edinburgh and Cambridge, every mathematician and physicist who met him saw past any behavioral awkwardnesses to his unique mind, one that combined an aptitude for abstract mathematics with a love of tinkering with homemade experiments. Maxwell was equally at home writing about topological geometry as he was making colored disks, which he would spin to demonstrate that red, green, and blue light combine to appear white.

The question of what heat is and why it flows from hot to cold hadn't interested Maxwell until he was at Aberdeen. It came to his attention because of his practical aspirations as a teacher. At Marischal College, the natural philosophy syllabus included heat, and to best teach this subject, Maxwell devised a set of experiments to carry out with his students. These included estimating the amount of heat needed to cause different substances to melt or boil. It was a successful endeavor.

Soon after he took up the post in Aberdeen, Maxwell started courting his boss's daughter. The principal of Marischal College, a Church of Scotland minister named Daniel Dewar, had been so impressed with the young man that he invited him several times to his home and then on a family holiday to the southwest coast of Scotland. Over these visits Maxwell formed a bond with Dewar's daughter, Katherine Mary, who at thirty-two was six years his senior. Their shared Christian faith helped cement their relationship, for James's surviving letters to Katherine from the time of their courtship are replete with long discussions of passages from the Bible. By February of 1858, the couple felt strongly enough to become engaged. In a letter Maxwell wrote to his aunt announcing the news, he says, "I can tell you that we are quite necessary to one another and understand each other better than most couples I have seen."

The phrase *quite necessary to one another* has an awkward ring about it, and unsurprisingly, the couple's life together was characterized by duty rather than affection. It was even alleged that Maxwell's first love was his cousin Lizzie Cay, but that the match was prevented for fear of consanguinity. Rather sadly, few of Maxwell's family and friends took to Katherine. As his cousin Jemima Blackburn wrote of her, "The lady was neither pretty, nor healthy nor agreeable, but much enamored of him. It was said that her sister had brought about the match by telling him how much she was in love with him, and being of a very tender disposition, he married her out of gratitude. Her mind afterward became unsettled but he was always most kind to her, and put up with it all. She alienated him from his friends and had a jealous streak." In later years, when the couple lived in Cambridge, people gossiped that Katherine disapproved when James socialized. There are apocryphal reports of her saying, "James, it's time you went home, you're beginning to enjoy yourself."

But the historical record of the Maxwell marriage is patchy. In 1929, a fire at the Maxwell ancestral home in Glenlair in the southwest of Scotland destroyed most of the family's papers, which may well have included letters that showed their marriage in a different light. And back in 1858, shortly after he proposed to Katherine, James wrote a poem declaring his love for her that contained this verse:

> Will you come along with me
> In the fresh spring-tide,
> My comforter to be
> Through the world so wide?

Over the years they stood by each other, nursing each other through near-fatal illnesses and occasionally collaborating in scientific research. Indeed, Katherine would make a crucial contribution to the advancement of the science of heat.

• • •

In February 1859 Maxwell came across Clausius's papers on the kinetic theory of heat. His initial response was that they were flawed. If he subjected them to rigorous mathematical analysis, he felt he would disprove them.

To do so, Maxwell turned to the laws of chance or, as they are more formally known, probability and statistics.

This was a radical step in the mid-nineteenth century. At that time, physicists saw it as their job to discover laws of nature that allowed predictions to be made with absolute certainty. They were in the business of saying what *will* happen, not what *might* happen. The shining example of such laws were Newton's laws of motion and his theory of gravity, which together made eminently reliable predictions of the orbits of planets, the trajectories of cannonballs, and other simple forms of movement. Yet Maxwell was suggesting that the way to test the kinetic theory of heat was to marry Newton's laws with the mathematics of chance, a subject widely believed to be outside the realm of fundamental physics.

In this, Maxwell was inspired by the work of astronomers. They didn't believe there was anything chancy or probabilistic about the behavior of celestial bodies, but they acknowledged that the human inability to make perfect observations blighted their work. If an astronomer measured the positions of a planet as it moved over time, the resulting measurements didn't show the true path of the planet. Despite the best intentions, each measurement had a small degree of error. The challenge was how to determine the true path of the planet from imperfect measurements. By the early nineteenth century, astronomers achieved this by adapting methods devised to assign odds in dice throwing games. When Maxwell was eighteen, he had read a paper describing this by the astronomer John Herschel, son of the discoverer of the planet Uranus. Their usefulness had struck the young man. As he wrote to a friend, "The true logic of this world is the Calculus of Probabilities. This branch of Math., which is generally thought to favour gambling, dicing and wagering, and therefore highly immoral is the only 'Mathematics for Practical Men.'"

To best explain this Calculus of Probabilities, it's instructive to go back to where the idea originated, in games of chance. Imagine a simple game in which if you correctly predict the outcome of flipping a coin, you win a bet. Assuming the coin is fair and not biased toward heads or tails, if you try to predict the outcome of one flip, there's a fifty-fifty chance you'll win.

And if you flip the coin a hundred times? Again, you will struggle to predict the outcome of any individual flip, but you will intuit that the chances of throwing a hundred heads and no tails is extremely unlikely. By way of contrast, the odds of throwing fifty heads and fifty tails are considerably higher. The exact odds are:

Odds of flipping exactly fifty heads and fifty tails: about one in twelve.

Odds of flipping zero heads and one hundred tails: about one in a million trillion trillion.

Few of us would back those odds. Unsurprisingly, the perfect fifty-fifty split is the most likely outcome, with the odds falling off precipitously as you move to more and more uneven splits of heads and tails.

If you graph the odds of each possible combination of outcomes of a hundred flips, you get a smooth mathematical curve, which looks like a cross-sectional slice through an old-fashioned church bell. (These graphs are often called *bell curves*.) The apex of the bell is halfway along the horizontal axis of the graph. This tells you that an even split of fifty heads and fifty tails is the most likely. As you move either left or right from this point, the graph slopes downward. This indicates that as the ratio of heads to tails grows more uneven, it becomes less likely. The far left of the graph represents the fact that you've thrown no heads. The far right—all heads. At these extremes, the curve is very close to zero.

The peak of the curve is the average number of heads thrown (fifty) over several sets of one hundred throws. A key feature of such curves is that overshooting the average has the same likelihood as undershooting it. So, throwing fifty-five heads has the same odds as throwing forty-five. These bell-shaped curves also require each item of data to be independent of all the others. Each coin flip or set of coin flips doesn't influence any of the others.

Bell curves are common in much scientific data as it often meets these conditions. One example is the heights of adults of the same gender—their heights will chart on a graph as a bell curve—as will their blood pressures. Or ask a competent marksman to fire one hundred times at a bull's-eye. Then count the number of bullet holes that landed within, say, one inch of the bull's-eye, the number that landed between one inch and two inches, two inches and three inches, and so on. Place these numbers on a graph and they will form a bell-shaped curve. (The better the marksman, the narrower the shape of the bell.) This process can be reversed. If the bullet holes form a telltale bell curve, you can estimate the location of the bull's-eye from the curve's peak. In a similar way, astronomers could deduce where a star really is from a series of inaccurate measurements of its position.

Sitting in his high-ceilinged granite office in Marischal College in the late 1850s, Maxwell applied the principle behind the bell curve to Clausius's ideas. The result would be a field of science known as statistical mechanics. Maxwell started by reiterating Clausius's claim that the temperature of a gas is proportional to the average speed of gas particles. But then Maxwell went in a new direction. He argued that some particles will move faster, and others slower, than the average, and these both play a role in the behavior of a gas.

But how to take account of them? Since there are around 10 million trillion particles in a cubic centimeter of gas, calculating the effect of each particle is impractical. So Maxwell introduced the laws of chance. Rather than concern himself with the speed of every single particle, instead he estimated the *percentage of particles*, in a given volume of gas, that might be moving at *any given range of speeds*. He argued that at any given temperature, there will be a speed at which the gas particles are most likely to move. However, there will also be particles moving faster and slower than this value. The chances of finding a particle moving at any given speed drops the further that speed is from the most likely value. The chances drop off in a similar way to the probability of getting more uneven numbers of heads and tails does in the coin-flipping example.

To see how this works, think of the nitrogen molecules that make up 78 percent of air around you. These are colliding into you all the time. Maxwell's analysis says that at room temperature, the most likely speed at which a nitrogen molecule will strike you is around four hundred and twenty meters per second. That's over nine hundred miles an hour. But for every hundred molecules that strike you at close to this speed, around fifty hit you at relatively sluggish speeds of around four hundred and fifty miles an hour. The same number hit you at a speedy fifteen hundred miles per hour. At the coldest temperature recorded on earth, around minus 90°C at the Vostok Station in Antarctica, the most likely molecular speed is down at about 740 miles an hour. However, for every hundred traveling at that speed, there will be nearly ninety moving at 900 miles an hour and around eighty moving at a mere 500 miles an hour.

A few years earlier, William Thomson had provided a definition of temperature at the macroscopic scale as a measure of the ability of heat to do work. Maxwell had now given temperature a definition at the microscopic scale in terms of the movements of tiny molecules.

Maxwell published his statistical analysis of how a gas behaves in early 1860 in a paper entitled "Illustrations of the Dynamical Theory of Gases." At the time it was a "speculation," to use Maxwell's own term. What turned it into a breakthrough was an observation that appears toward the paper's end. Maxwell found that his statistical analysis made a prediction about how a real gas behaves that was testable by experiment. This prediction was unique to kinetic theory and thus a means to verify or disprove it.

The prediction didn't directly deal with heat. It concerned the viscosity or stickiness of a gas. We don't think of gases being sticky in the way a liquid such as honey is, but they are, simply less so. Hold your hand out flat and move it slowly through air—a small viscous frictional force acts to resist the movement.

For kinetic theory's explanation of this phenomenon, picture a thin metal plate, above and parallel to the ground. It moves very slowly through air at a constant speed. The air between the plate and the ground consists of trillions of tiny billiard ball–shaped particles in constant motion. The ones in contact with the moving plate are dragged along with it. As you move farther down and away from the plate, the plate's presence becomes less noticeable. The particles' tendency to move in the direction of the plate falls. Instead you notice the effect of the ground, which is to slow down the particles. And at the ground, you find the air particles directly in contact with it are stationary.

But according to kinetic theory, air particles move randomly in all directions, not just in the same direction as the plate. This means some of the faster moving particles near the plate move downward. They then collide with slower ones in the lower air, causing these to speed up slightly. And similarly, slower particles from near the ground move up and collide with the higher, faster ones, slowing them down.

The cumulative effect of all these collisions manifests as the resistive force on the plate.

Maxwell's mathematical analysis of kinetic theory made a counterintuitive prediction about this force. It said that it does not change as the gas's pressure varies. Even though reducing the pressure means there are fewer air particles between plate and ground, the air will not be less sticky.

This happens because reducing the number of air particles has two effects, which cancel out. On the one hand, there are fewer collisions between the particles as they move up and down. This lowers the overall resistive force on the plate. But on the other hand, *because* each particle has fewer other ones to bump into, it moves a greater distance between collisions. Therefore, its reach is greater. It can slow down particles that are farther away. The slow particles near the ground exert their "slowing-down" force more effectively. Thus, the resistance the moving plate feels remains the same.

Conversely, in higher pressure air, there are more particles per unit volume. But they travel much shorter distances between collisions. The sheer number of particles shields the moving plate from their ability to slow it down. (This calculation assumes the air's temperature remains unchanged and thus the speeds of its constituent molecules stay the same.)

The importance of this prediction that pressure does not affect viscosity is that it is a necessary consequence of kinetic theory. If experiments didn't bear it out, then the whole theory and everything it had to say about heat and temperature would collapse. So Maxwell searched the scientific literature to see if any data existed on the relationship between a gas's viscosity and its pressure. What little there was seemed to run counter to his prediction. A few lines below the mathematical equation in question, he writes, "The only experiment I have met with on the subject does not seem to confirm it."

Could it be that, as Maxwell had suspected, kinetic theory was untrue? The existing data was too patchy and inconclusive. And it's apparent from a letter he wrote that in applying himself to the theory he'd grown attached to it: "I am getting quite fond of it and require to be snubbed a little by experiments."

Maxwell decided the only option was to design and perform an experiment that measured the viscosity of a gas as its pressure varies. But before he could carry it out, circumstances intervened.

A few short months after he published his paper, Maxwell lost his job. Aberdeen, despite being a small town, had two universities, Marischal and King's, and in 1860, the city authorities decided to cut costs by merging them. They subsequently decided the new institution couldn't afford two professors of natural philosophy. Unfortunately for Maxwell, his opposite number at King's, David Thomson, a six-foot-six giant of a man with a domineering personality to match his physique, was the prime mover behind the merger. Also, Maxwell was the younger of the two. This ensured that he was the one to be let go.

Further misfortune struck. In the autumn of 1860, while on a visit to his family home of Glenlair in the southwest of Scotland, Maxwell contracted smallpox. For a month, he battled this devastating disease, which killed three out of every ten people who caught it. Fortunately for physics, James survived. He later told friends that Katherine's nursing night after night saved his life. In later years when Katherine underwent prolonged bouts of illness, James never stinted in his duty to look after her in turn.

Then in October 1860 the Maxwells moved to London. James had won a new post as professor of applied sciences at one of Britain's newest universities, King's College London, founded in 1829. He and Katherine took up lodgings in a terraced house in Kensington, a borough near Hyde Park in the capital.

Here Maxwell returned to the question he had been investigating in Aberdeen. Although Katherine Maxwell had shown little interest in the mathematical aspects of her husband's work, by the early 1860s she

Maxwell's apparatus to measure
gas viscosity

had developed a passion and skill for the experimental side of physics. Together, in the garret of the Kensington terrace house, the Maxwells built a makeshift laboratory, where one of the problems they tackled was verifying the kinetic theory of heat by testing Maxwell's prediction that the viscosity of gas is unaffected by its pressure.

The apparatus the Maxwells assembled to do this was ingenious yet simple—resembling the kind of stove you might find in a log cabin. It was essentially a four-foot-long, thin, hollow brass pipe that sat on top of a glass bowl. Suspended parallel to the ground inside the glass bowl by wires that ran from the top of the brass pipe were seven thin metal disks, three that moved and four that were fixed. By using magnets held below the glass bowl, the Maxwells could make the movable disks twist back and forth. They filled the whole apparatus, including the brass pipe and the glass bowl, with air whose pressure they could measure with gauges attached to the brass pipe.

The Maxwells used magnets to twist the disks and allowed them to freely oscillate. Next they measured how long a single twist or oscillation of the disks took at different air pressures. If Maxwell's mathematics was right then the disks would oscillate at the same rate irrespective of the air's pressure.

The Maxwells measured the twists or oscillations of the disks with an artful trick. They attached a mirror to the wires with which the disks were suspended and shone a ray of light onto it. As the disks twisted back and forth, so did the mirror. This caused the ray of light that reflected off it to sweep across a piece of ruled paper stuck on the wall some six feet away. The tiny movements of the mirror, thus enlarged, could be measured with great accuracy.

The Maxwells spent several months in their Kensington garret making meticulous measurements with this apparatus. It wasn't easy. In addition to taking readings for air pressure and meticulously timing the disks' oscillations, they also had to ensure that the temperature of the gas inside the apparatus didn't vary, which required a large fire to be continuously stoked for hours on end, even during the hot summer months.

The results, when they eventually came, were a triumph for Maxwell's mathematics and kinetic theory. The measurements showed that through a wide range of air pressures, from so low that the air only exerted a pressure equivalent to half an inch of mercury all the way up to so high that it exerted a pressure equivalent to thirty inches of mercury, the disks took the same amount of time to oscillate back and forth. The viscosity of the air remained the same throughout the entire range of pressures.

By experimentally verifying a prediction made only by kinetic theory, Maxwell gave humanity a plausible explanation for what heat is and why things feel hot or cold. The kinetic theory of heat allows us to visualize what happens in the world we inhabit at scales too small to see—namely that everything around us consists of tiny particles in constant motion, and how hot or cold something feels is simply the way we, at a macroscopic scale, sense that motion.

Though today Maxwell is best known for his work on electricity and magnetism, in the 1860s his contemporaries were well aware of his papers on kinetic theory. When he was spotted in a jostling crowd leaving the Royal Institution's public lecture hall by no less a figure than Michael Faraday, Faraday drew the analogy of members of the crowd colliding with one another to particles doing the same in a gas, saying, "Ho, Maxwell, cannot you get out? If any man can find his way through a crowd, it should be you."

But for all its impressive descriptive power of heat, kinetic theory still fell short in one crucial way. It still didn't explain why heat spontaneously flows from hot to cold objects. This discovery had been one of the great achievements of early nineteenth-century science and was now enshrined as a universal law of nature, namely the second law of thermodynamics. Yet nothing that Maxwell had said about kinetic theory shed light on why it was true.

It's somewhat perplexing that Maxwell didn't connect the dots and extend his own statistical analysis to explain the second law. After all, he'd introduced statistics to physics and, with Katherine, carried out a crucial experiment to establish the validity of this approach. From his writings, it's clear that his intuition was telling him that the second law had some connection to statistics. But his attention had switched away from gas theory and thermodynamics to electromagnetism. For much of the 1860s, he focused his intellectual energy on the latter, eventually delivering his seminal mathematical analysis of the subject in 1873. That analysis not only described all electric and magnetic phenomena, but it would reveal the true nature of light, enable the invention of radio, and inspire Einstein's work on relativity.

In addition, in 1871, Maxwell was appointed the first head of Cambridge University's new physics laboratory, the Cavendish. Teaching now became his main preoccupation. In this institution, the next five generations of scientists would discover the electron and the neutron, split the atom, and uncover the structure of DNA. Maxwell threw himself into the task of creating the laboratory, supervising both the design of the building and the apparatus it would contain. Then, sadly, abdominal cancer, the same disease that had killed his mother, struck Maxwell. He died in 1879, aged forty-eight. Katherine stayed on in the family home in southwest Scotland, living in obscurity, until she died seven years later.

By the early 1860s, kinetic theory was widely accepted. But the second law of thermodynamics remained as mysterious as ever. Physicists could say why a cup of tea felt hot, but not why, when left to its own devices, it cooled down.

# CHAPTER TEN

# Counting the Ways

Mathematics is a language.
—Josiah Willard Gibbs

The staccato opening chords of Beethoven's *Eroica* symphony sounded like artillery fire as they ricocheted around the auditorium of the Vienna Philharmonic. It was the summer of 1866, and among the audience was a bearded, bespectacled twenty-two-year-old named Ludwig Boltzmann. Of less than average height and sporting a mop of curly dark hair, he was a PhD student at Vienna University's physics department. A gifted pianist from childhood, Boltzmann understood how Beethoven had grabbed Western classical music by the scruff of its neck and dragged it in a completely new direction. What the young man didn't yet know was that over the next four decades, in a career that would match that *Eroica's* turbulent swings of mood and tempo, he would do the same for physics.

At the same time, another man from another continent, Josiah Willard Gibbs, would embark on an equally important lifelong investigation into the mysteries of thermodynamics. For in 1866, while Boltzmann was soaking up Viennese culture and writing his PhD thesis, the twenty-seven-year-old Josiah Willard Gibbs was heading eastbound on a steamer across the Atlantic from America to begin a three-year tour of Europe's great cities. This was Gibbs's first and only trip outside his native New England. While in Europe, he attended lectures in science and mathematics, acquiring the intellectual tools necessary for his later work on energy and entropy.

While Gibbs's and Boltzmann's scientific interests overlapped, they were, in every other sense, opposites. Where Gibbs was a thin and reclusive ascetic, Boltzmann was chubby, gregarious, and passionate, prone to mood swings that took him from exuberance to despair. If the Austrian's life is characterized by Beethoven's *Eroica*, the American's might be represented by one of Erik Satie's austere musical contemplations. And though the two men both took the laws of thermodynamics as their starting point, they set off in different directions. While Boltzmann looked inward, seeking to understand why these laws must be true, Gibbs looked outward, seeking to understand their consequences.

Ludwig Boltzmann was born in Vienna on February 20, 1844. That year the date fell on Shrove Tuesday, which in the Christian calendar is traditionally celebrated as the last festive day before the austerity of Lent begins. In later years Boltzmann would joke that this explained why his moods switched from great happiness to deep depression in an instant. Boltzmann's father was a tax official with the official title "regional finance commissar" in the Hapsburg government, and his mother was the daughter of an affluent merchant from Salzburg. Ludwig was an able student who regularly came top of his class and displayed both a curiosity about the natural world and a talent for music. He collected butterflies and beetles and took piano lessons from the great composer Anton Bruckner, who at the time was a cathedral organist. Though Ludwig had, as a future colleague put it, "stubby fingers and pudgy hands," they proved no handicap to his skills as a pianist.

In the nineteenth century, however, affluence offered little protection from disease. In 1859, when Boltzmann was fifteen, his father died of tuberculosis, and within a year his younger brother, Albert, succumbed to the same illness. The family could get by on Boltzmann's father's state pension and his mother's inherited money for a while, but not indefinitely. In consequence, Boltzmann soon became the sole provider for his mother and sister. A career as an academic scientist at a university would suit his talents and could provide financial stability, but were such careers available in mid-nineteenth-century Austria?

The Industrial Revolution had come later to Austria than to Prussia and the other German-speaking states to the north, as had the Hapsburg government's realization that the study of science was important for a modern state. Had Boltzmann been born two decades earlier, there would have been nowhere in Austria for him to pursue a career as a paid physicist.

Fortunately for him, in 1850, the government had agreed to fund a Physics Institute at Vienna University. It consisted of only two or three teaching staff and fewer than twenty students, whose main reason for being there was to be trained as secondary school teachers. They occupied cramped and makeshift quarters in a part of Vienna called Erdberg, which is near where the river Danube divides the city.

What the Vienna Physics Institute lacked in facilities and materials, however, it made up for with a staff with a strong sense of camaraderie and a passion for scientific research. Years later Boltzmann would remember his time here as the happiest of his scientific career.

The elder statesman of this small coterie of scientists in Vienna was Josef Loschmidt, a lecturer twenty-three years older than Boltzmann. The older man mentored the young, fatherless Boltzmann, and an enduring friendship developed based on a shared passion for science, art, poetry, and music. The pair frequented Vienna's theaters and concert halls and attended the 1866 performance of Beethoven's *Eroica*. They blurred the boundary between art and science, discussing Homer and the art of the Sistine Chapel within the physics department and the properties of sulfur crystals while queuing for the opera. These outings, judging by Boltzmann's later writings and the rate at which he put on weight, also involved the consumption of large amounts of beer, wine, and food. (His wife, Henriette, would call Boltzmann "my sweet fat darling.")

An admirer of Rudolf Clausius's and James Clerk Maxwell's papers on kinetic theory, Loschmidt had used their ideas to deduce the diameter of a single particle of air. His figure of a millionth of a millimeter is the first such estimate—it's about three times the modern value for the diameter of oxygen and nitrogen molecules, which make up air. Importantly, Loschmidt introduced kinetic theory to Boltzmann, who was entranced, especially by Maxwell's work. If Beethoven was Boltzmann's artistic hero, Maxwell was now the scientific equivalent—the Scotsman's science had a similar effect on his emotions as the German's music. As he wrote of Maxwell's papers on kinetic theory:

"Ever higher surges the chaos of formulae. Suddenly, four words sound out: 'Put N = 5.' The evil demon V vanishes, just as in music a disruptive figure in the bass abruptly falls silent, that which had seemed insuperable has been overcome as if by a stroke of magic."

Completing what Maxwell had started by introducing the laws of chance to physics would define Boltzmann's career. Using statistics to explain the second law of thermodynamics, to explain why the entropy of the universe always increases, would become the obsessive focus of his work, the white whale to his Ahab.

All that was to come. After receiving his PhD in physics from Vienna University, Boltzmann's chief concern was money. He worked for a while as assistant to his head of department, Josef Stefan, but this didn't pay well enough to support his family. Fortunately, Stefan, by now well aware of Boltzmann's scientific talent, wrote him an effusive letter of recommendation. With this, Boltzmann obtained a post as chair of mathematical physics in the town of Graz, Austria's second-biggest city, situated about 120 miles southwest of Vienna. In September 1869, the twenty-five-year-old Boltzmann and his family took up residence in the town.

Graz had been a thriving melting pot of German, Italian, and Slavic peoples since medieval times, and it had a venerable university, founded in 1585. Physics, however, had only been taught there since 1850. The new department was even humbler than that in Vienna, consisting of a priest's residence that had been converted into a makeshift laboratory and a small lecture theater. As the laboratory was unheated, the man who ran the department, Professor Toepler, whose previous post had been at the Baltic town of Riga, lent Boltzmann a thick fur coat so that he could work through the winter months. In these unpromising circumstances, often bitterly cold, Boltzmann launched his lifelong scientific campaign to unravel the mysteries of heat.

Boltzmann's first significant work was a paper published in 1872 in Austria's leading scientific journal. Entitled "Further Studies in the Thermal Equilibrium of Gas Molecules," it was lengthy and repetitious—and audacious. Its core argument was that the second law of thermodynamics, namely that the entropy of the universe always increases, was a direct result of kinetic theory.

To see how Boltzmann combined these two ideas, imagine an oven in a large kitchen. Once the oven is hot, switch off the power and open its door. Heat always moves from hot to cold, as Clausius and Thomson had shown, so the temperature of the air inside the oven will cool until it's the same as that of the room. But why? This is the question Boltzmann sought to answer. First, remember that at the instant the power is switched off, according to kinetic theory, the air inside the oven feels considerably hotter than the air outside because, on average, the air molecules inside are moving much faster than those in the air outside. Now picture what happens at the oven's open doorway where the hot and the cold air meet. Here, by chance, fast-moving particles from inside the oven will collide, from time to time, with slow-moving particles from the outside. Boltzmann believed that the solution to the mystery of the second law of thermodynamics would be found in these collisions

Working out how trillions of such collisions unfold over time, however, was an immense mathematical challenge. Boltzmann's friend and mentor Josef Loschmidt had shown that individual air particles are tiny, and that a cubic centimeter of air contains around 10 million trillion. Calculating exactly what occurs as each collides with another seems impossible.

Boltzmann's approach to the problem was original and brilliant. He knew that a fast particle possesses more energy of movement—kinetic energy—than a slow one. Think here of the kinetic energy of a moving object as a measure of how much effort is required to bring it up to its speed of travel from rest. Or, equivalently, how much "braking effort" it would take to stop it. The faster and heavier an object, the greater the effort required in both cases, and the greater, therefore, its kinetic energy.

Kinetic energy is a useful concept when analyzing collisions. For instance, picture a moving cue ball approaching a stationary ball on a billiard table. Some of the cue ball's kinetic energy is lost as frictional heat as it rolls on the table, some turns into the sound of the collision with the stationary ball, and some is transferred to the latter, which now moves. The cue ball, having given up a portion of its kinetic energy to the ball it has struck, then travels increasingly slowly until it stops. What Boltzmann did was to visualize gas particles as idealized billiard balls. Except when they collide, no kinetic energy is lost in sound or friction; it's only transferred from the more energetic particle to the less energetic one.

The air in a room consists of a vast number of particles that swap kinetic energy back and forth. That's how Boltzmann pictured it. To simplify his analysis, he employed a mathematical trick. He divided the amount of kinetic energy that a single particle carries into discrete whole-number units. So, a particle might move with 1 unit or 6 units or 35 units of kinetic energy, but never with 2.3 or 5.78 or any non-integer number of units.

This simplified Boltzmann's calculations, while providing a realistic description of actual molecular-energy transfers. This technique also enabled a way of visualizing molecular behavior. To see why, imagine that instead of large numbers of air molecules moving with different energies, there are crowds of jostling people with differing numbers of coins in their pockets. Individuals in the crowd move in random directions but only advance by one or two paces before bumping into another person. In this analogy, a fast molecule with many energy units is represented by a person with a large number of coins, a slow one by a person with a small number. A hot oven in a cool room can thus be seen as a small group of rich people bunched together in the corner of a large room full of much poorer ones.

Continuing the analogy, the equivalent of the temperature of each of the two groups is the average wealth of its members. Some individuals in each group are richer or poorer than the average, just as with gases of two different temperatures some particles are moving quicker or slower than the average. The equivalent of a fast molecule bumping into a slow molecule and losing some of its energy is a rich person bumping into a poor person and handing over some of his coins, making the rich person poorer and the poor person richer. With these rules in mind, follow the money.

At first, only the rich at the edge of their group will become poorer because they are the most likely to bump into members of the surrounding poor. The poor at the edge of their group will become a little richer, thanks to these same collisions. It's not possible to take account of every such encounter because there are so many, but it is possible to predict how the distribution of the coins throughout the crowd will change over time.

Eventually, the transfers of money that mainly took place at the "rich-poor" border will start to spread. Rich people farther from the rich-poor border will also start losing money because their neighbors on the border are no longer as rich as they once were. Similarly, the poor on the border will soon lose their gains to their neighbors deeper in the "poor hinterland." Soon all the money that was concentrated in the hands of the rich will be spread among the poor.

To cement the thought experiment, make the numbers of people involved smaller. Say twelve people are in a room. A block of six on the left have one coin each, and a block of six on the right have nothing. Add a further simplification—each person can only hold one coin at a time. The coins randomly move around the room as people swap them or give them to a person without one.

How will the coins end up being distributed?

To answer, count the number of ways the coins can be distributed that look similar to one another. So, in this example, all the distributions with six on the left and none on the right look identical because the coins are identical. What about distributions with five on the left and one on the right? These aren't identical but will look like one another. Similarly, all the distributions with four coins on the left and two on the right will resemble one another. And so on.

Now ask how many ways can all six coins be held by the people on the left? There's quite a few. The first person could hold any of six coins, the second person any of five, and so on. So, the total number of ways of distributing all the coins on the left is 6 × 5 × 4 × 3 × 2 × 1, which equals 720.

How many ways are there for five coins to be anywhere on the left and one to be anywhere on the right? The answer is considerably bigger: 4,320.

For four on the left and two on the right? 10,800.

For three and three? 14,400. The ways of achieving this even spread outnumber those for all the other distributions.

For two on the left and four on the right? 10,800.

For one on the left and five on the right? 4,320.

For none on the left and six on the right? 720.

Even with this small number of coins, more spread-out distributions

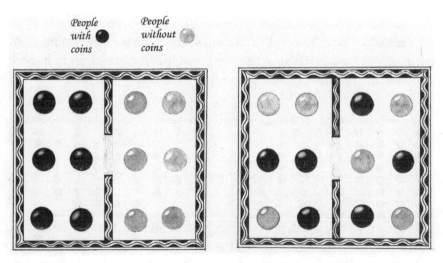

The distribution on the left is twenty times less likely to occur than one like that on the right, where equal numbers of coins are on each side of the room.

such as four on the left and two on the right, three on the left and three on the right, and two on the left and four on the right are more numerous than uneven spreads. Glance into the room after a thousand random coin exchanges have occurred and there's a 31 percent chance you'll see three coins in the left half of the room and three on the right. By way of contrast, there's only a 1.5 percent chance that all of them will be on the left. If the system starts off with all the coins held by people on the left, over time the money will tend to spread out.

With larger numbers, this effect is even more pronounced. Say the room contained one hundred people and fifty coins can move about. Distributions in which the coins are equally or nearly equally spread vastly outnumber uneven distributions by a factor of several billion.

Note that each arrangement, whether spread out or uneven, is by itself very unlikely. But because all the many trillions of spread-out arrangements are *indistinguishable from one another*, it's almost certain that the coins will end up in one of these.

Boltzmann applied this same logic to the way heat dissipates. The only difference is, instead of people swapping coins, molecules are transferring kinetic energy.

In essence, Boltzmann showed that there are far more indistinguishable ways for small numbers of energy units to be distributed throughout the kitchen than there are for large numbers of energy units to be concentrated among a small number of molecules. Any system that starts off in an unusual or rare type of distribution, such as the room when much of the heat was concentrated in the oven, will inevitably end up in a more common distribution, i.e., with heat dissipated and spread out.

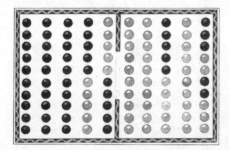 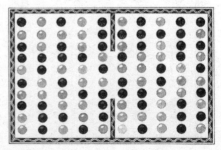

With more coins, uneven distributions like the one on the left are vastly outnumbered by evenly spread-out distributions like the one on the right.

Or, put another way, heat will always dissipate away from a hot region because after a period of random collisions the odds are stacked overwhelmingly in favor of that result.

Entropy, by Boltzmann's reasoning, is simply the number of *indistinguishable* ways the constituent parts of a system can be arranged. To say entropy increases in any given system is another way of saying that any given system evolves into ever-more-likely distributions or configurations. The second law of thermodynamics is true for the same reason that when a pack of cards arranged in suits is shuffled, it will end up jumbled. There are many more indistinguishable ways for the pack to be disordered than there are for it to end up ordered, and so shuffling takes it in that direction.

This "number of distributions" method of defining entropy extends its explanatory power far beyond the dissipation of heat. It readily explains many irreversible processes in nature. For example, air rushes out of an opened balloon, and never back in, because there are many more ways for the air particles to be spread out throughout the room than for them to be crammed inside the balloon. Similarly, there's no way to stir milk in a cup of tea so it separates instead of mixing because there are many more ways for the milk particles to spread throughout the tea than to be concentrated in one spot. Likewise, dropping an egg breaks it and creates a mess, but if we take the mess from a broken egg and drop that, it never reforms itself back into the egg. Again, that's because the number of ways the scattered egg particles can form an unbroken, whole egg is far outnumbered by those in which it remains a mess.

Entropy increases with time, therefore, because the chances of its decreasing are tiny. In fact, and this is the mind-blowing part of Boltzmann's logic, only by observing entropy increase can we tell the direction of time. We differentiate the future from the past because in the future the overall entropy is greater. So, by trying to understand heat in terms of atoms, Boltzmann had uncovered what lay behind William Thomson's discovery of time's arrow. Imagine you're watching a film that shows the heat in a kitchen flowing back into an oven, or a film showing the milk in a teacup separating back out. You would rightly surmise that the film was running backward. Time's arrow is simply a reflection of the inexorable march from statistically unlikely ordered arrangements to more likely disordered ones. There's a subtle point here: a movie showing heat flowing back into an oven isn't showing something that couldn't happen; it's showing something that's extremely unlikely to happen. But it's so unlikely that, if we see it, we know something is awry.

Boltzmann's 1872 paper had flaws but it is nonetheless a scientific landmark, marking the first serious attempt to provide an explanation for the second law of thermodynamics at the molecular level. At the time, though, it made little impact. This was partly because the circle of professional physicists in Austria was rather small, and no one there felt able to comment on Boltzmann's complicated mathematical arguments. Germany was more promising in this regard. When Boltzmann visited Berlin in 1872, the professor of physics at the city's university, Hermann Helmholtz, did take an interest in his ideas. But little came of it; Boltzmann, used to the informal academic culture of Austria in which professor and student socialized, found it hard to speak freely with Helmholtz, the embodiment of the more formal, hierarchical nature of Prussian society. "Not so accessible," is how Boltzmann described the eminent physicist in a letter to his mother. Prussia's universities were more prestigious than their Austrian counterparts, but the gregarious Boltzmann found them unwelcoming and bureaucratic. Not for the first time, he struggled to have his ideas heard.

By 1872, the same year that Boltzmann was considering the underwhelming response to his paper, Josiah Willard Gibbs, back in his birthplace of Yale, was also hard at work. Gibbs, however, assumed nothing about the structure of matter. His strategy was to look away from the molecular causes underlying the laws of thermodynamics and, instead, to consider their consequences.

Gibbs came from an intellectual tradition. His father, also named Josiah Willard, had been professor of sacred literature at Yale and an accomplished linguist. An active abolitionist, Gibbs senior had played an important role freeing enslaved Africans who had rebelled aboard the Spanish ship *La Amistad*. How he did so prefigures his son's approach to science. What happened here was that in the summer of 1839, fifty-three captives from Mendeland, now in modern-day Sierra Leone, broke free a few days after their ship sailed out of Havana. They took control of the vessel and demanded that her navigator sail them back to their homeland in Africa. The navigator, however, misled the Africans and instead took *La Amistad* to North America. Here, the US navy commandeered the vessel and the Africans were interned in New London, Connecticut. It was now up to the US justice system to decide—were the Mende captives on *La Amistad* the property of their Spanish owners, or were they free people who had rebelled in self-defense?

The problem faced by American abolitionists, for whom *La Amistad* became a cause célèbre, was that the Mende spoke no language comprehensible to the inhabitants of New England. With no means for the Mende to tell their side of the story, mounting their legal defense would be well nigh impossible. Willard Gibbs Sr. determined to solve this problem. He visited the Mende in their Connecticut jail, and when he met them, he held up first one finger, then a second, and, one by one, all the fingers of his hands. The Mende understood his intent and told him their words for the numbers from one to ten. Gibbs then traveled to New York harbor and went from ship to ship repeating the words for the numbers that he had learned from the Mende. Eventually he found a sailor, a freed slave working on board a British brig, who understood the Mende numbers and who spoke English. Thanks to the universal nature of numbers, Gibbs had found the interpreter the captives in Connecticut needed. Their legal challenge for freedom could begin, and though the process took over two years, the US Supreme Court ruled that the Mende had been illegally held and transported as slaves, and their rebellion was an act of self-defense. It ordered them freed, and thirty five survivors returned to Africa.

Although Josiah Willard Gibbs Jr. was only two then, he grew up with the lesson that mathematics is a universal language. In time Gibbs would use it to extend the power and relevance of the laws of thermodynamics far beyond its origins.

As with many of his predecessors, Gibbs chose to study thermodynamics because he couldn't ignore the transformative power of steam-driven technology. By the mid-nineteenth century, America had witnessed a frenzy of railway building. Indeed, the American Civil War, which was at its height in 1863 when Gibbs was searching for a topic for his PhD, was the first conflict in history in which rail had dominated the logistics of war, both in moving and supplying troops. The North's eventual victory was due in part to their superior steam technologies, just as Britain's victory over France in the Napoleonic Wars was due in part to theirs. Gibbs's PhD thesis was "On the Shape of Teeth in Spur Gearing." He followed that with a patent application for a brake for railway cars.

Applied science, however, would not be Gibbs's path. When his father,

Gibbs senior, died in 1861, he left his children a substantial estate, including bonds in three Midwestern railway companies. This inheritance provided Gibbs and his two sisters the means to undertake their three-year tour of Europe. Gibbs used the trip to expand his scientific horizons. Though he did not officially enroll at a European university, he attended lectures on subjects ranging from the abstract mathematics of number theory to the physics of light, sound, and heat. In Heidelberg, Gibbs sat in classes given by Hermann Helmholtz, who had pioneered the principle of the conservation of energy.

When Gibbs returned to America, he took up an unpaid position as professor of mathematical physics at Yale. The university offered no salary, and thankfully, Gibbs did not need one. With a comfortable house and ample funds from his father, Josiah Willard Gibbs was ready for his labors of the next decade.

The fathers of thermodynamics, Carnot, Joule, Thomson, and Clausius, had seen their field as a way of understanding the relationship between heat and work. Gibbs freed it from this narrow constraint. He showed how all behavior in the material world, from the way solids melt and liquids boil to the way chemical reactions proceed, all obey the laws of thermodynamics.

Initially, however, Gibbs's scientific ambitions were modest. His stated aim was to make these recently discovered laws easier to comprehend. He worried especially about the concept of entropy. *He* could understand Thomson's and Clausius's definition that it is a measure of how dispersed heat is throughout any substance, but could anyone else? "The notion of entropy . . . will doubtless seem to many far-fetched, and may repel beginners," he wrote perceptively.

In 1873, in his first two scientific papers, Gibbs addressed this problem in a way that was simple and consequential. He drew maps. In the same way that geographical maps give us an immediate understanding of a landscape, Gibbs pioneered the idea of thermodynamic maps—charts that show how the physical properties of a substance change as, for example, it's heated or cooled, squeezed or stretched. These reveal the laws of thermodynamics at work in the material world.

For example, heat water in a saucepan on a stove while measuring its temperature. You're at sea level and the water starts off at room temperature of 20°C. This rises steadily as heat flows into the water.

Then boiling starts. The saucepan contains a mixture of water and steam. You make an important observation—the temperature has stopped rising. It stays fixed at 100°C. Heat continues to flow in from the stove, but its effect is to make steam, not make the water hotter.

Then, only when all the water has turned into steam, the temperature rises again.

Now, imagine doing the identical experiment in La Rinconada, Peru, the highest town on earth. The altitude is fifty-one hundred meters, and the atmospheric pressure there is about half that at sea level. You note two differences from when doing the experiment at the first location: the water boils much cooler at 83°C, and it remains as a water-steam mixture for longer. Darwin noted this effect while camping in the Andes. Despite potatoes being boiled all night, they weren't soft enough to eat.

Now picture the inside of a pressure cooker. The sealed lid means, as the water boils, the steam increases the downward pressure on the water to double that of the earth's atmosphere. You find now that the boiling point of water has risen to 121°C, but it exists as a water-steam mixture for less time.

Graphically, the three experiments look like this (where the lines are horizontal, the water is boiling and there's a mix of water and steam in the saucepan):

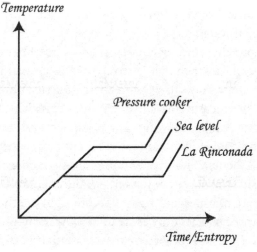

How water temperature changes
as it is heated

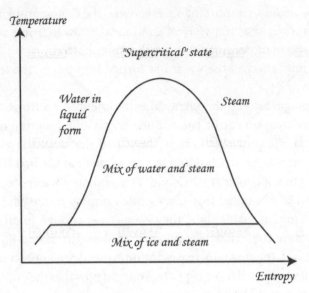

A thermodynamic map of water and its phases

Remember the definition of entropy—it's a measure of how dispersed heat is through any object—in this case the water in the saucepan.

So, the time axis in this case also represents the entropy of the water and the steam. Heat from the burning gas has steadily been entering the water, dispersing through it.

This chart shows that entropy increases can manifest in two ways. Either as a rise in temperature *or* as a change from water to steam. In the latter case, the increasing entropy shows itself as the increasing steaminess of the water.

By repeating this measurement over a range of pressures, we obtain our first thermodynamic map, showing how water, ice, and steam respond to being heated and cooled in a wide range of circumstances. To the left of the dome are all the temperatures and pressures where water exists in its liquid form. Inside the dome are the temperatures and pressures where it exists as a mixture of water and steam. To the right are the temperatures and pressures where it exists only as steam. Below the dome, the pressures and temperatures are so low, water exists as a mixture of ice and steam. Above the dome, the temperature and the pressure are so high, water is neither liquid nor vapor; it's in what's called a supercritical state.

The importance of these charts cannot be overstated. They are widely used, for instance, by engineers who design the type of power stations that generate much of the world's electricity. Many of these contain modern-day steam engines in which heat from coal, nuclear reactions, geothermal sources, or sunlight is used to create hot, high-pressure steam. Unlike in their nineteenth-century counterparts, this doesn't push a piston. Instead, it rushes through turbine blades, making them spin and drive electricity generators. After the steam has done its work spinning the turbine, it's condensed back into water and the whole process repeats. The overriding concern here is efficiency—to convert as much of the heat available into electrical power. Thanks to Sadi Carnot, engineers know the best way to achieve this is make the steam as hot as possible. But they must do this while maintaining the structural integrity of the power station's component parts.

This is where thermodynamic charts are invaluable. They tell engineers, for example, precisely how much heat energy is absorbed by the power plant's water as it turns into steam and how much pressure it exerts and how hot it gets. The charts tell the engineers the best temperature at which to condense the steam exiting the turbines back into water. This allows them to maximize the efficiency of the power station while keeping it safe.

Turn on a light, watch TV, roast a chicken in an electric oven, a thermo-dynamic chart is a big part of making that possible.

But Gibbs's work extends far beyond power generation. His map-based or graphical approach focused scientists' and engineers' minds on so-called *phase changes*, the term representing a transition of a material from, say, liquid to gas or solid to liquid and vice versa. A thermodynamic chart reveals these to be transitions at a constant temperature at which the entropy of a material changes dramatically. In other words, paradoxically, during phase transitions materials can absorb heat without getting hotter and can reject heat without getting colder. We saw with water that while it's boiling, its temperature stays fixed at 100°C. Similarly, if you cool steam that's at, say, 120°C, once this drops down to 100°C, it starts to condense back into water. But then the temperature will stay fixed until all the steam is converted into water. Only then will the temperature fall again. Understanding phase changes enabled humans to master the ability to make things cold.

Archaeology suggests our ancestors figured out how to make fire around a million years ago. Making ice was much harder, and the power to cool is the unglamorous but indispensable technology of the modern age. Refrigeration is the most obviously thermodynamic of all human inventions, and the most defiant of the universal tendency for entropy to increase. These devices force heat to pass from a cold interior to a warm exterior, which is in the opposite direction to the one in which heat flows spontaneously. The purpose of this is to create a space where the relentless increase of entropy is slowed down. Although ostensibly a refrigerator is a cool box, that's a means to an end. Its ultimate purpose is to slow down decay and putrefaction, which are both examples of entropy increasing. Think of a refrigerator as a device inside which time slows down.

It's hard to overstate how important this has been to human progress and well-being. By greatly extending the time for which food can be safely stored and transported, the refrigerator enabled the greatest improvement in human nutrition since our prehistoric ancestors discovered the pathogen-killing benefits of cooking. We eat safer, healthier food now more than ever before in our history, and in addition, refrigeration is vital to the widespread use of vaccines, which has saved millions from premature death and disease. The steam engine is much credited for being the invention that catalyzed the Industrial Revolution, but the refrigerator isn't far behind in importance.

The preservative benefits of keeping food cold have long been known, but not until the early nineteenth century did "ice harvesting" become a global business. Frederic Tudor, the Boston "Ice King," became one of postrevolutionary America's first millionaires, shipping ice cut in New England to the Caribbean, Europe, and even India. At its peak, the US ice trade employed an estimated ninety thousand people. Norway exported a million tons of ice a year, drawing on a network of man-made lakes.

But could ice be made artificially? Many tried, hoping to use the fact that evaporating liquids have a cooling effect. It's why we sweat, after all. Many, such as Benjamin Franklin, had noticed in the 1750s that when a cleaning fluid, diethyl ether, evaporates at room temperature, it cools its surroundings much more dramatically than water.

A century passed till machines appeared that exploited this effect on a commercial scale. One, designed by a Scottish emigrant to Australia named James Harrison, who clearly felt the heat in his adopted country, could make several tons of ice per day. These devices, the ancestors of modern refrigerators, used steam power to pump liquid ether through a coil of tubes that enclosed a chamber containing water. This turned into ice as the ether in the coil evaporated.

Breweries bought these early refrigeration units, once again making an honorable appearance in the history of physics. Lager-style beers are fermented at just above the freezing point of water, and plentiful artificial ice meant they could even be brewed in hot summers. Then in the 1870s, around the time Josiah Willard Gibbs was writing his papers on thermo-dynamics, the first ships fitted with mechanical refrigeration, "reefers," began carrying frozen meat and poultry across the Atlantic. Undertak-ers started using artificial cooling to prevent corpses from decomposing.

As with steam engines, the early pioneers of making things cold didn't focus on the physics underlying their machines. But also as with steam, that changed. Physics and engineering aligned as was embodied by a Ger-man scientist, engineer, and entrepreneur named Carl Linde. Born in Bavaria in 1842, he studied engineering at the Swiss Federal Institute of Technology, where his teachers included one of the founding fathers of thermodynamics, Rudolf Clausius. Linde then moved to Munich, where he became professor of engineering at the Technical High School. (One of his students there was Rudolf Diesel.)

Linde focused his talents and his understanding of thermodynamics on refrigeration. By 1875, he was using thermodynamic charts of the kind pioneered by Gibbs to improve the effectiveness of refrigerators. In prin-ciple, these charts can depict the behavior of any substance as it's heated or cooled, and Linde experimented with chemicals such as ammonia with great success. In 1879, he left academia and set up the Linde Ice Machine Company, in the town of Wiesbaden. His machines were considerably better than his competitors', and in a decade he sold 12,000 units in his native Germany and around 750 in America, mainly to breweries. Then in 1892, Guinness in Dublin asked Linde if he could supply them with liquid carbon dioxide, to improve their beer's foamy head. This inspired him to study the liquefaction of air, which Linde succeeded in doing on a large scale—reaching temperatures below −140°C. That in turn enabled the industrial production of pure oxygen and nitrogen. The widespread electrification of homes in the early twentieth century then made domes-tic refrigerators a reality.

Refrigerators, particularly in the domestic setting, exploit the physics of phase changes. Their coolants are chosen from volatile substances that boil at a low temperature of around 4°C. Behind the refrigerator's inner wall is a network of pipes called an evaporator. The coolant evaporates inside it, sucking heat out of the device's interior at a constant temperature of 4°C. But then, given that heat never flows spontaneously from cold to hot, how do you get the newly formed coolant gas, which is at a low temperature of 4°, to release its heat into surroundings that are much warmer, typically by 20°?

The answer is that the coolant is forced through a device called a compressor, which is the opposite of a cylinder in a steam engine. In those, heat turns into work by making a gas expand. In a compressor, work is turned into heat by squeezing a gas. Once enough heat is added in this way to ensure the coolant vapor's temperature is greater than that of the room, it is allowed into the condenser, a network of pipes at the back of the refrigerator. Here the vapor releases into the surroundings the heat that was in the refrigerator's interior *and* the heat that was created in the compressor. Put your hand behind your refrigerator to feel that combined heat.

In the condenser, as it releases heat, another phase change occurs as the coolant turns back into a liquid. But this is quite warm, it's the same temperature as the room. For the refrigeration process to continue, the coolant's temperature must fall back to 4°C before it can reenter the evaporator. To achieve this, the coolant liquid passes through a tiny nozzle called an expansion valve. As the coolant is forced through it, its pressure drops, and it cools and it's ready to enter the evaporator once again.

The compressor ensures the refrigerator complies with the second law of thermodynamics. Heat flows out of the refrigerator interior lowering its entropy. But the total heat that flows out of the condenser raises the room's entropy to compensate. We thus pay for a small entropy-reducing space in our kitchens by speeding up the pace at which the entropy of the universe increases.

Back in 1873, Gibbs had little idea of the epic consequences of his papers. Modest and unassuming, he sent his work to the little-known *Transactions of the Connecticut Academy of Arts and Science*, which had no readership outside Yale. Moreover, because Gibbs's papers were longer than the articles usually published by the *Transactions* and because they contained mathematical formulas, their typesetting costs exceeded the publication's budget. To cover these, the editorial committee had to obtain donations from other faculty members and local businessmen. One committee member, A. E. Verrill, later recalled that they had long discussions about the merits of Gibbs's papers even though no one on the committee understood them. "Yet we all believed what Gibbs wrote must be of intrinsic value in his branch of science. Therefore, we raised the money and printed each paper as it came in."

# CHAPTER ELEVEN

# "The Terroristic Nimbus"

I am conscious of being only an individual struggling weakly against the stream of time.

—Ludwig Boltzmann

In Graz, while Gibbs was writing his papers, Ludwig Boltzmann found someone with whom he could discuss his ideas. In May of 1873, he met the nineteen-year-old Henriette von Aigentler while she was training to be a schoolteacher. Ten years younger than him, with long fair hair and blue eyes, she formed a friendship with him in part because of their shared interest in science. The year before she met Boltzmann, even though Austria's universities did not permit women to take degrees, von Aigentler had attended physics lectures at Graz University. After her first term, the university excluded all female students on the grounds that their presence distracted the men present. Undeterred, von Aigentler petitioned the Education Ministry and obtained character references from sympathetic lecturers. This persistence enabled her to continue attending lectures for another term.

Surviving correspondence from the early years of von Aigentler and Boltzmann's relationship suggest that she took the initiative. When her mother died, leaving her the youngest of three parentless sisters, her letters became more frequent, though they still mainly discuss her studies and his scientific career. The first sign of romance is when von Aigentler asked Boltzmann for a photograph of him as a memento. Boltzmann did comply, but his replies to her many chatty letters are brief. Then in September 1875, von Aigentler's persistence paid off when Boltzmann sent her a rather formal marriage proposal, which ended with the plea that a wife be "a comrade in a shared endeavor."

Von Aigentler, however, could not break free of nineteenth-century misogyny. Once the couple announced their engagement, family pressure compelled her to give up even her unofficial scientific studies and, instead, learn cookery. She found this frustrating and complained to Boltzmann in early 1876, "Unfortunately, I have right now very little time to read or study, since I am sometimes even in the evening in the kitchen of Kienzl."

Over their thirty-year marriage, Boltzmann would often need von Aigentler's support because, in contrast to his expectations, his ideas came in for repeated criticism, which would exacerbate a psychological predisposition to depression and despair.

The first and most constructive criticism Boltzmann received was from his friend and mentor Josef Loschmidt. The latter's motivation was that he disliked the way the second law of thermodynamics predicted that the universe would eventually die, degenerating into a never-changing state in which all heat had dissipated throughout the cosmos. If this was true, wrote Loschmidt, the second law is a "terroristic nimbus cloud, which appears to be a destructive principle to all life in the universe."

To counter this bleak vision, Loschmidt used a subtle argument, pointing out an apparent paradox in Boltzmann's reasoning.

To follow Loschmidt's reasoning, think again of the spontaneous flow of heat from the inside of a hot oven whose door is open, to a large room. This is a typical irreversible process, for once switched off, an oven always cools down till the temperature of both oven and room are the same. And the reverse never happens of its own accord.

Boltzmann had explained this as a natural consequence of countless collisions between air molecules that individually follow the same rules of physics that billiard balls do when they collide. Therein, claimed Loschmidt, lay the paradox. The laws that describe each individual molecular collision are reversible. They are completely symmetrical in time. To see why, imagine watching a film that shows a close-up collision between two billiard balls. One ball enters from the left and strikes a stationary one. The first ball stops, and the second ball exits to the right. Now imagine what you'd see if the film is reversed. A ball would enter from the right, strike a stationary ball, and stop, and the stationary ball would exit to the left. It's impossible to tell in which case the film is running forward in time and in which, backward.

Now apply this principle to the example of heat dissipating out of an oven. Imagine a camera that zoomed in and filmed just one of the trillions of collisions occurring between the air particles at the open door of the oven. By watching the film, you couldn't tell if it was running forward or backward in time. But if, subsequently, the camera zoomed out and filmed the whole room, then you could tell which way the film was running. If it showed heat spreading outward from the oven, you would conclude it was running forward in time. If it showed the opposite, heat leaving the room and spontaneously concentrating itself into the oven, you would say the film had been reversed.

According to Loschmidt, Boltzmann was claiming that an irreversible process such as the dissipation of heat is the result of many individual reversible collisions, and this is a paradox. For how can an irreversible result arise from a reversible process? It just doesn't make sense. Where does that irreversibility come from? Boltzmann was an astute enough scientist to acknowledge a valid criticism, and he agreed that, at the microscopic scale, individual molecular collisions are reversible. So, he wrote two papers, published in 1877, which broadened his arguments and strengthened the proposition that entropy goes up for statistical reasons alone.

Especially in the second paper, Boltzmann used dense mathematics to make his case as rigorous as possible. "Elegance is for the tailor and the shoemaker" is how he justified this approach, which was markedly different from that of his hero James Clerk Maxwell. Toward the end of the paper, Boltzmann formally defined the idea that entropy arises for statistical reasons alone with the following equation:

$$\Omega = -\int\int\int\int\int\int f(x,y,z,u,v,w)\ln f(x,y,z,u,v,w)dxdydzdudvdw$$

In later years, Boltzmann's intellectual successors, by the adroit use of symbols, shortened his original to:

$$S = k\ln W$$

Now considered one of the foundational statements of physics, this formulation is inscribed on Boltzmann's grave in Vienna. It's a mathematical statement that means the entropy (S) of any system is the number of indistinguishable arrangements it can take.

Meanwhile, Josiah Willard Gibbs had also been busy in his Yale study. He'd realized that the laws of thermodynamics could bring a deep, new understanding to the field of chemistry. Above all else, Gibbs would give future generations of scientists a framework for understanding the chemical processes that occur within living organisms. These ideas appeared in his magnum opus, a 371-page epic paper, packed with mathematical symbols and thus testing again the fundraising skills of the editorial committee of the *Transactions of the Connecticut Academy of Arts and Science*. That James Maxwell had, on reading Gibbs's first papers, added a whole chapter describing their thermodynamic maps in a new edition of his book *Theory of Heat* encouraged the editorial committee's continued faith in Gibbs.

Gibbs's insight was to find a way showing how the two laws of thermodynamics drive all chemical reactions. He chose to start his argument with a restatement of those laws, so let's follow his lead:

First law: The energy in the universe is constant.

Second law: The entropy of the universe tends to increase.

Gibbs then showed how all processes of change can be judged by these two laws. He did this, essentially, by turning the two laws into one new law we can call Gibbs's law:

The flow of energy is the means by which the entropy of the universe is increased.

First, let's remind ourselves what we mean by a chemical process or reaction. The simplest explanation is that a chemical reaction describes what happens when substances combine with others to form a new substance. Take iron rusting, for example. Iron combines with oxygen and water vapor to form a new material—rust. When baking soda combines with vinegar, it forms carbon dioxide, water, and salt. Soap removes grease by a chemical reaction in which the two combine to form a new substance that is soluble in water. Cooking is replete with examples, as is every living body. Gibbs's law allows us to understand why every single chemical reaction takes place.

Consider an everyday chemical reaction such as coal burning in a grate. In this process, carbon, the main constituent of coal, combines with oxygen in the air to form carbon dioxide, giving off a great deal of heat. (Most coal has impurities as well that also react with oxygen, but for the purposes of this illustration, let's ignore those.) Why don't we ever see this process in reverse?—the film running backward and the carbon dioxide turning spontaneously back into coal and oxygen? Why can't you put the heat that was given off during burning back into carbon dioxide and separate it back into solid carbon and oxygen?

The answer lies in Gibbs's idea that energy will always flow to increase the entropy of the universe. Consider what happens when coal burns.

To start with, we have solid carbon and gaseous oxygen. For an intuitive sense of the entropy of this situation, think of the energy as more densely packed in the solid carbon and more dispersed in the gas.

After burning, there's only the gas, the carbon dioxide. The energy that was concentrated in the solid carbon has become more dispersed. What had started off as a mixture of a low-entropy solid and a high-entropy gas has turned completely into a high-entropy gas. Overall, the entropy of the materials has gone up.

And, important, as carbon and oxygen combine, heat is released and flows into its surroundings, into the air around the grate, thus warming it up. This heat then disperses through the air, causing its entropy to go up.

The reason carbon burns but carbon dioxide never spontaneously "unburns" is that the burning causes a twofold increase in entropy. First it creates carbon dioxide gas, and second it disperses heat through the air around the grate. All told, it's an effective way of increasing the entropy of the universe.

Burning coal like this is analogous to the example of the house with one hot and one cold room, which we met when discussing entropy in chapter 7. To strengthen the analogy, imagine the door between the rooms is held shut by a spring. To start with, nothing happens, just as nothing happens to coal if it sits undisturbed in a grate. Then a hand reaches in and opens the door. Heat starts to flow. The hand disappears but the door remains open. A small proportion of the heat flow is turned into the mechanical work needed to keep it so. The mysterious hand is equivalent to the spark needed to initially ignite the coal. This energy needed to kick-start a reaction is usually termed the activation energy. But once the coal is burning, ample heat is generated for the process to continue.

We don't observe carbon dioxide being unburned for the same reason that we don't observe heat flowing spontaneously from the cold to the warm room. Neither process would contravene the first law of thermodynamics—no energy would be destroyed or created—but they would reduce the entropy of the universe, which is forbidden by the second law. All such reactions that result in an increase in entropy are called spontaneous. That means they will proceed as long they have received the activation energy required to start them.

Another example of a spontaneous reaction is when hydrogen burns in oxygen to create steam, the gaseous form of water. To start with, there are two distinct gases, which is a fairly high entropy state. Energy is dispersed through two gases. After burning, there's only one gas—steam. Two gases have become one, which is a fall in entropy. But the burning produces a great deal of heat, which disperses into the surroundings, causing a large increase in the entropy there—much larger than the fall due to the two gases becoming one. Overall, the entropy of the whole system has gone up. Just as with carbon dioxide, water never unburns by itself into its constituent gases because that would require a fall in entropy.

Water never unburns *by itself*. Gibbs's law stipulates that although the entropy of the universe must go up, the entropy of its component parts can go down. This can happen as long as the entropy of other parts of the universe go up by enough to ensure that the sum total of the entropy in the universe has increased.

In other words, it is possible for carbon dioxide and water to "unburn"—plants do it all the time—but not "by itself." Gibbs's equation allows us to tot up all the entropy changes in different parts of the universe to reveal a marketplace—one in which one bit of the universe pays other bits of the universe for a highly desirable commodity—a local and temporary reduction in entropy. And it does so with a specific and well-defined currency—energy.

Imagine two houses, each of which has two rooms. In the first house, the door between the rooms has been replaced with an engine. In the second, it's been replaced with a refrigerator, which pumps heat the wrong way from the cool room to the hot one. The flow of heat from hot to cold in the first house drives the engine, creating work. This work, in turn, drives the refrigerator in the second house.

In effect, the flow of heat from hot to cold in the first house has powered a flow of heat the "wrong" way, from cold *to* hot, in the second house. The increase in the entropy of the first house has bought, using the currency of work, a decrease in entropy in the second. The two houses are said to be "coupled."

Chemical reactions can be coupled in the same way as the rooms in this diagram.

When hydrogen burns in oxygen, a great deal of heat is dispersed, far more than is needed to compensate for the fall in entropy caused by

*House with engine*

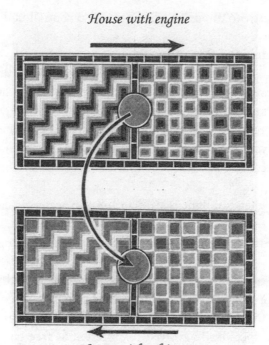

*House with refrigerator*

Thermodynamically "coupled" houses

creating the steam. This excess or "free" energy can be used to produce mechanical work—such as drive a car engine. But it can also be used to force other chemical reactions to proceed in the reverse or "nonspontaneous" direction, just as heat flow in one house reversed it in the other. In this context, the available energy is often referred to as Gibbs free energy, and it's the means by which chemical reactions are coupled.

For instance, under the right circumstances, the free energy left over from burning hydrogen and oxygen can unburn carbon dioxide. The former reaction increases the entropy of the universe and the latter decreases it. As long as the quantities are such that the overall entropy goes up, then the burning of one can lead to the unburning of the other.

That Gibbs free energy can couple one chemical reaction to another enables all life on earth. The most spectacular example of this is the very first step in the process—photosynthesis, which is essentially the use of

Gibbs free energy to unburn both water and carbon dioxide. The process works as follows:

Step 1: Capture the free energy in sunlight.

Sunlight is an abundant source of free energy. The chlorophyll molecule in the leaves of plants uses this to unburn water—in other words, to split the $H_2O$ molecule into its constituent parts of hydrogen and oxygen. The oxygen is released into the atmosphere, leaving hydrogen, on its own, within the leaves. Isolated hydrogen, of this kind, is itself now a source of free energy because it needs to rebond with oxygen or anything chemically similar to oxygen.

This step, when sunlight is used to unburn water, is called the light reaction.

Step 2: Use the free energy in the form of isolated hydrogen to unburn carbon.

The clever aspect of this part of the process is that the plants don't release the free energy stored as isolated hydrogen in one go. Instead, they divide up the free energy held in the isolated hydrogen into other chemicals that have specifically evolved to store Gibbs free energy. The most common of these is adenosine triphosphate or ATP. Think of ATP as a tiny molecular spring that becomes coiled when it receives free energy. That packet of energy can then be accessed on demand by the chemical equivalent of releasing the spring in the ATP.

Using the Gibbs free energy stored in ATP, plants unburn carbon dioxide. In a series of choreographed chemical reactions, the free energy in each ATP molecule is released and used to split the carbon and oxygen in atmospheric carbon dioxide and repackage them into molecules known as carbohydrates. This is known as "fixing" carbon and it has two main purposes. First, carbohydrates provide building block materials such as cellulose, which give the plant structure. Second, making carbohydrates doesn't use up all the energy stored in the ATP. The unused energy is, in effect, transferred into the carbohydrate molecules. They are also chemical springs. This means that carbohydrates themselves become temporary stores of free energy, which can then be used to fuel plant growth and all the other chemical reactions a plant needs to live.

This second step in photosynthesis, when the free energy in isolated hydrogen is used to fix carbon, is called the dark reaction.

This is what makes animals such as us possible. From a Gibbs free energy perspective, we're plants that run backward. When we eat plants or eat other animals that eat plants, we're ingesting chemicals such as carbohydrates that plants have created and turned into rich stores of Gibbs free energy. In a precise reversal of the dark reactions in photosynthesis, animal cells release the free energy stored in carbohydrates and use it to make their own ATP molecules. These fuel many of the chemical processes that take place in animal cells, enabling them to live. At the end of this process, the carbon that the plants fixed from atmospheric carbon dioxide is reunited with oxygen and breathed out again as carbon dioxide.

So, in summary, plants use the Gibbs free energy in sunlight to turn water and carbon dioxide into carbohydrates that contain some of the original solar free energy, while releasing oxygen. Animals access the Gibbs free energy trapped in carbohydrates to live and, by so doing, recombine the carbon in the carbohydrates with atmospheric oxygen to emit carbon dioxide and water. Scientists have now accounted for every single transfer of Gibbs free energy in all the chemical processes that occur in plants and animals and overall. A beautiful symmetry is at work here. Plants take in 2,870 kilojoules of solar free energy to make 180 grams of glucose (a typical carbohydrate). An animal that eats 180 grams of glucose releases exactly 2,870 kilojoules of free energy, eventually breathing out carbon dioxide.

This is the famous cycle of life. Carbon dioxide breathed out by animals is absorbed by plants, which make food and oxygen and so on. The cycle needs a constant supply of Gibbs free energy to turn. And crucially, at each step of the cycle, a small amount of free energy is lost as heat. This means at each step the entropy of the universe goes up. The cycle of life, for all its glory and wonder, exists between sunshine and a sewer. Overall, life is an effective way of increasing the entropy of the universe.

Gibbs himself didn't investigate the role of free energy in biology. But within fifty years of his paper's publication, the mysteries of photosynthesis and its reverse counterpart in animals, cell respiration, were unraveled. Knowing, from Gibbs's work, and the work of those who followed, that the transfers of free energy were the driving force behind life gave biochemists a guiding principle to follow as they unpicked the devilishly complicated details of cell chemistry.

Gibbs's ideas were also significant in the long-standing debate over vitalism, the theory that living organisms were governed by different scientific principles than inanimate things. The work of Hermann Helmholtz and others had done much to weaken support for vitalism, but Gibbs's work was a more decisive blow. The notion of free energy showed that every chemical process in every cell in every living creature fell inside the realm of physics. Nothing supernatural or spiritual was needed. The energy in rays of sunlight was enough to power the intricate beauty of life on earth.

In the years following the publication of Gibbs's magnum opus in 1878, his ideas slowly took hold in the scientific community. That he made no assumptions about the structure of matter or molecules helped him to avoid controversy. The same, unfortunately, could not be said for Boltzmann. As the last decades of the nineteenth century progressed, his scientific ideas embroiled him in a row that damaged his already fragile psyche. However unintentionally, Gibbs's work played a part in that.

The fight originated with a philosophy known as phenomenalism, which had taken hold in the German-speaking world. Its main proponent was Ernst Mach, the professor of the history and philosophy of the inductive sciences at Vienna University.

In his younger days, Mach had been a gifted experimental physicist, taking the first photographs of the shock waves created by objects that travel faster than the speed of sound, which is why that speed is now known as Mach 1. By the 1890s, phenomenalism had become his passion. At its simplest, phenomenalism is the view that only things that can be sensed directly can be said to be real. Conversely, attributing physical reality to something for which there is only indirect evidence is therefore poor science. For Mach, the problem with Boltzmann's work was that it assumed the reality of atoms and molecules. Given that neither could be sensed directly, phenomenalism argued that it could not be said that they existed. That in turn meant that the entire edifice of statistical mechanics that Boltzmann had built was suspect. His probabilistic explanation for the way entropy increases was worthless if the objects it depended on, namely atoms and molecules, could not be observed directly.

Many scientists in the German-speaking world were attracted to phenomenalism at that time, including the young Albert Einstein. In later years he recalled how Mach's ideas made him realize that time and space have no meaning unless there are clocks and rulers to measure them, and that in turn helped inspire his theories of relativity. But the cost for

Boltzmann was high. He found himself drawn into exhausting debates over the existence of the atoms and the molecules that he claimed were in constant, violent motion.

Applying phenomenalism to thermodynamics also meant discussing the subject only in terms that could be observed and measured—heat flows, pressures, volumes, temperatures, and so on. This view became known as energeticism. Its proponents held that although the notion of atoms and molecules may well have enabled some fancy mathematics, that did not make them real. Here they turned to the work of Josiah Willard Gibbs.

The point is that Boltzmann's ideas were predicated on a world built of atoms and molecules, whereas Gibbs had made no such assumptions. Instead Gibbs had created a comprehensive and rigorous description of thermodynamics based solely on its two well-accepted laws. This pure approach appealed to the energeticists. One, a young German chemist named Wilhelm Ostwald, even translated Gibbs's papers into German. Excellent though this was for science, as Ostwald helped raise awareness of Gibbs's work in Europe, it was less good for Boltzmann, as it meant the energeticists could, with some justification, say that thermodynamics had no need for the hypothesis that atoms and molecules are real.

Among Boltzmann's critics was a young physics lecturer in Munich named Max Planck. Born in Kiel on Germany's Baltic coast in 1858, Planck had received his PhD for an analysis of the second law of thermodynamics. Like Mach and Ostwald, he disapproved of Boltzmann's insistence on the reality of atoms and molecules and that their behavior underpinned the second law. In 1882, he explicitly declared Boltzmann was wrong, writing, "The second law of the mechanical theory of heat is incompatible with the assumption of finite atoms. . . . A variety of present signs seems to me to indicate that atomic theory, despite its great successes, will ultimately have to be abandoned."

The battle between Boltzmann and the energeticists played out across journals and in scientific conferences. Witnesses recalled exhausting debates between Boltzmann and Ostwald with neither changing the other's view. On one occasion, after a lecture Boltzmann gave at the prestigious Imperial Academy of Science in Vienna, Mach stood up and declared, "I don't believe that atoms exist!"

Mach's words, Boltzmann later said, "ran around in my head." By then, he had dedicated over twenty years of his life to thinking about the scientific consequences of our universe being built from atoms and molecules. He'd bet his entire career on their reality. Now approaching the end of that career, he found a new generation of scientists were drifting away from him. In a letter to the editor of a scientific journal in the 1890s, he wrote, "Whether I will soon be alone in opposing the present direction of German science I cannot say."

The double frustration was that this move away from his ideas was motivated not by mathematical arguments, nor by physical evidence, but by philosophical musings, which Boltzmann found ultimately pointless. "Shouldn't the irresistible urge to philosophize be compared to the vomiting caused by migraines?" he asked in a letter to Italian philosopher Franz Brentano.

Boltzmann's health had also begun to deteriorate. His myopia was so bad that he struggled to continue his laboratory work, and friends noticed how his joie de vivre was punctuated by ever-longer periods of silent withdrawal. Then in 1889, tragedy struck when his oldest son died suddenly of appendicitis, and Boltzmann blamed himself for not spotting the symptoms earlier. University life also became a challenge. Nationalism was on the rise, and Austrian students seemed in perpetual conflict over whether they should be for or against Germany, a debate that Boltzmann found as baffling as it was exhausting. These rows occasionally degenerated into drunken riots. In later years, he described these as a group of pigs whose tails curled to the left fighting another group whose tails curled to the right.

Worse, even scientists who weren't ardent phenomenologists and energeticists began to find fault with Boltzmann's work. When the mathematician Ernst Zermelo published a critique in 1896, for instance, Boltzmann began his defense, "Zermelo's paper shows that my writings have been misunderstood; nevertheless it pleases me for it seems to be the first indication that these writings have been paid any attention in Germany."

Nonetheless, Boltzmann was still capable of original thought. His rebuttal of Zermelo contained many remarkable ideas, the most striking of which is the first statement from scientific reasoning alone that the universe must have originated in a single moment of creation.

## CHAPTER TWELVE

# Boltzmann Brains

I sleep badly and am quite beside myself with misery. . . . Please forgive me everything!

—Ludwig Boltzmann

In 1854, William Thomson had concluded from observing heat dissipate through an iron bar that the universe must die. Four decades later, spurred by criticism, Ludwig Boltzmann would argue that the statistical explanation of entropy implied that the observable universe must have been born. This was decades before astronomers discovered evidence for the big bang. And though Boltzmann's version of creation differs from the modern one, elements of it are crucial to the study of cosmology today.

The notion that the universe has a moment of creation came from Loschmidt's and Zermelo's criticisms, which forced Boltzmann to admit that lurking in his statistical explanation for the second law of thermodynamics was an unstated assumption. Boltzmann had claimed that the increasing entropy of the universe is a result of its moving from less to more likely configurations. This explanation only works, conceded Boltzmann, if you assume that the universe started off in a very statistically unlikely low-entropy state.

To see why, imagine a jar containing several black and white marbles. Say the marbles are mixed to start with. Then as the jar is shaken, the marbles move from one mixed state to another. Film of that would look the same running forward or backward. However, to use the jar of marbles as a way of revealing time's arrow, you have to first arrange them in layers—one of black and one of white and so on—and only then shake the jar. Now if the film shows the marbles becoming more mixed, you know it's running forward in time. The marbles are going from less to more likely arrangements. Applying that principle to the universe, it implies that at any instant in the past, the universe must have been in a more "unlikely" state than it is in now. The further back you look, the more unlikely the configuration the universe was in. And that begs the question, How did the universe get into an unlikely, low-entropy state in the past?

Boltzmann's answer was that there had to have been a moment of creation. No god was required, only natural phenomena and the laws of chance.

The universe as a whole, Boltzmann hypothesized, is in a state of never-changing equilibrium. Imagine it is a vast, featureless cloud of gas. Nothing happens other than the random collisions of gas particles. For all intents and purposes, this universe is dead. But by chance alone, after eons of inactivity, a small part of the universe deviates from this state and fluctuates accidentally into an unusually low-entropy state. Stars and galaxies pop into being in this part of the universe by chance. It's like saying that if you shake the jar of black and white marbles long enough, by chance alone they will become ordered into neat layers. Continue shaking, however, and they will lose that order and return to being jumbled. The small section of the universe we inhabit is rather like that, Boltzmann argued. Once, long ago and by chance, it fluctuated into a low-entropy state, and since then, by chance, its entropy has slowly been increasing, which will eventually return it to equilibrium with the rest of the dead universe. But because it's only possible for life to exist in this low-entropy segment of the universe, living beings observe time's one-way arrow. Or as Boltzmann put it:

"A living being that finds itself in such a world at a certain period of time can define the time direction as going from less probable to more probable states (the former will be the 'past' and the latter the 'future'), and by virtue of this definition he will find the universe is 'initially' always in an improbable state."

Most of the universe is dead, but we can't know about that dead part because it's impossible to live in. This kind of reasoning is known as the anthropic principle, which states that a universe that humans live in must run according to physical laws that permit human life. This sounds like a tautology, but these days physicists and cosmologists often use this principle to explain the mysterious fact that the universe appears "fine-tuned" in a way that's conducive to our existence. For instance, the strength of gravity, the mass of the atomic nuclei, the speed of light, and the other "constants of nature" are precisely the values they need to be for the universe to be stable for billions of years. If any of these values were even slightly different, the universe would either collapse in on itself or all the stars would burn out within seconds. One way of explaining this is the idea that we live in just one universe within a "multiverse." This consists of many other universes where the constants of nature are indeed unsuitable for life. But because we can't live in those universes, we are only aware of our universe, with its values for the constants of nature. Although the context of this reasoning is rather different than Boltzmann's, the method, the anthropic principle, is his.

So how plausible is Boltzmann's creation story—that our habitable universe began as a random, highly unlikely fluctuation that created a region of low entropy? Most modern cosmologists would probably reject Boltzmann's explanation, but simply by raising the question, Boltzmann defined a topic at the heart of modern theoretical physics. To see why, we have to unpick the key flaw in his "random fluctuation" hypothesis—the argument against it goes as follows:

Our universe is complicated and highly structured—it's not only capable of life but is also full of stars and galaxies, all of which are individually very ordered systems. That means that the entropy of our universe is remarkably low. And that means that when it began, it was even lower, in a supremely ordered state of super-low entropy. In other words, our universe is highly unlikely. So far there's no contradiction with Boltzmann's suggestion. All we have to do is wait long enough and such an unlikely event will eventually occur.

But consider the following: If random fluctuations are what bring a universe into existence, we can easily imagine a fluctuation that's much more statistically likely than the one needed to bring our entire universe into existence. An example of such a fluctuation is one that brings only a solar system just like ours into existence. Beyond it, there would be nothing. Nonetheless, life could exist in this lonely solar system. We could exist in it. In fact, it's much more likely that a solar system surrounded by a featureless dead universe came into existence than what we see, which is our solar system surrounded by billions of complex galaxies.

Keep going on that logical path. A fluctuation that brings one habitable planet into existence is much more likely than one that brought an entire solar system into being. From there, it's just a short step to arguing that a fluctuation that brings only the room I'm sitting in into existence with nothing outside but a featureless universe is even more likely than one that brings only the planet into existence.

Take that line of reasoning to its extreme conclusion, and you end up with the idea that a fluctuation that brings only a single brain into existence is far more statistically likely than any of the above.

Boltzmann's explanation for the original low-entropy state of our universe therefore leads to the solipsistic conclusion that all that exists is one brain, which holds the entire cosmos as a figment of its imagination. Scientists now refer to this hypothetical entity as a Boltzmann brain. Few favor such an explanation, although no one has yet come up with a definitive explanation for why the universe did begin in such an unlikely low-entropy state. As the great American physicist Richard Feynman, speaking in the 1950s during his now-famous lectures on physics, put it:

"For some reason, the universe at one time had a very low entropy for its energy content, and since then the entropy has increased. So that is the way toward the future. That is the origin of all irreversibility, that is what makes the processes of growth and decay." But then he concludes, "The one-way behavior of the entire universe—it cannot be completely understood until the mystery of the beginnings of the history of the universe is reduced still further from speculation to scientific understanding."

It's been over half a century since Feynman's talk, and over a century since Boltzmann first proposed the low-entropy origins of the universe. The open question of how that occurred is now a vibrant research field.

Yet at the turn of the twentieth century, Boltzmann had no idea that his scientific successors would revere him. Caught in the psychological torment of defending his ideas, feeling under attack, his physical health worsened as he struggled with asthma and increasing corpulence. "Papa sweats and swears all the time," his son Arthur wrote. This combined with the attacks by Mach and his allies sapped Boltzmann's confidence. In an 1898 letter to his student Felix Klein, he wrote, "Just when I received your dear letter, I had another neurasthenic attack."

*Neurasthenic attacks* were a common fin de siècle term for bouts of anxiety or depression. When they worsened, Boltzmann was forced to seek psychiatric help and endure stints in a sanatorium in the countryside near the town of Leipzig in Germany. Nothing helped. In a letter in 1900 to his wife, Henriette, he wrote, "I sleep badly and am quite beside myself with misery. . . . Please forgive me everything!"

Unaware of Boltzmann's suffering and of the way his own work had boosted the Austrian's critics, Gibbs was cautiously coming around to accepting that there was something to the atomic and molecular hypothesis. In 1902, Gibbs published a textbook entitled *Elementary Principles in Statistical Mechanics Developed with Special Reference to Rational Foundations of Thermodynamics*, which contains similar arguments to Boltzmann's without wholeheartedly adopting the latter's belief in atoms. Gibbs was too careful a scientist to nail his flag to that mast, writing, "Certainly, one is building on an insecure foundation, who rests his work on hypotheses concerning the constitution of matter."

In April 1903, Gibbs, who unlike Boltzmann had lived a life of equanimity, suddenly suffered an acute intestinal obstruction. The doctors were unable to help and Gibbs died at home, alone, just as he had lived his life.

Two years later, in the summer of 1905, Boltzmann had one last happy period in life when he accepted an invitation to lecture at the new universities of Berkeley and Stanford in California. His account of the trip, "Journey of a German Professor to Eldorado," reveals his love of the novelty and the energy of America, noting, "Whenever I enter the harbor of New York, I am seized by a kind of ecstasy."

From New York, Boltzmann took a four-day train journey west to Berkeley. There he was struck by admiration for the way wealthy individuals such as railway magnate Leland Stanford and Phoebe Hearst—the wife of a mining millionaire whose son, William Randolph Hearst, would be the inspiration for *Citizen Kane*—poured money into academic institutions.

Boltzmann had only two complaints about America. The first was the food. Staying at Phoebe Hearst's opulent hacienda, he was baffled by a serving of oatmeal, "an indescribable paste of oat flour with which one might perhaps feed geese: but I am not sure, certainly a Viennese goose would not eat it." The worse problem was that large parts of the country, Berkeley included, were under the grip of temperance movements and so were dry. The lack of alcohol, Boltzmann claimed, caused him dreadful dyspepsia.

When Boltzmann returned home, he was in the best of spirits. "California is beautiful, Mount Shasta magnificent, Yellowstone Park wonderful," he wrote in his journal, "but, by far the nicest part of the whole journey is the moment one is back home again."

Yet within a few months, the depressive forces that had always lived somewhere in Boltzmann's mind reasserted themselves. Vienna was little comfort. Coming home seemed to have awakened his frustrations with the energeticists. One Vienna University student, the great physicist Lise Meitner, who in the 1930s would play a crucial role in the discovery of nuclear fission, recalled that despite being a passionate and entertaining lecturer, Boltzmann struggled to cope with these attacks on his ideas. Some fifty years after the event, Boltzmann's lecture was still a vivid memory:

"The lecture was really a most stimulating experience. . . . Boltzmann had no inhibitions while he spoke, . . . he told us how much difficulty and opposition he had encountered because he had been convinced of the real existence of atoms."

Then in September 1906, Boltzmann, his wife, and his daughter went on holiday for a few days to the Italian coastal town of Duino, near Trieste, in northeastern Italy. One day his wife and daughter went down to the beach to bathe, leaving Boltzmann alone in their rooms. When Boltzmann's daughter returned from the beach, she found her father had hanged himself.

# CHAPTER THIRTEEN

# Quanta

I was ready to sacrifice any of my previous convictions about
physics.

—Max Planck

In 1900, Max Planck, a critic of Boltzmann's science for nearly two decades,
published papers that hinted at a change of heart. Even more unexpectedly he seemed to be saying that Boltzmann's statistical methods might
have relevance far beyond thermodynamics.

This reluctant conversion was forced upon Planck by the advent of a
new technology—the electric light bulb. In these electric current flows
through a filament, warming it and making it glow. This focused scientific
minds on investigating the precise relationship between heat and light.

There are three ways—conduction, convection, and radiation—that
heat can flow out of an object. All can be observed in most kitchens.

Conduction is how electric hot plates transfer heat. The whole heated
surface of the plate is in contact with the underside of a pan, and the heat
flows from one to the other. Kinetic theory explains this as follows: As the
hot plate's temperature rises, its constituent molecules vibrate at faster and
faster rates. Because they're touching the molecules of the saucepan, they
shake them. Soon all the saucepan molecules are vibrating more vigorously than before, which manifests as the saucepan's temperature rising.

Heat flow through convection occurs in ovens. The heating elements
within the oven's wall cause the air molecules nearby to zip about more
quickly. These then collide with molecules deeper in the oven, increasing
their speed, and soon the entire oven's temperature rises.

The third kind of heat transfer, by radiation, is the one linked to light.

Turn on a grill, and as the element's temperature rises, it glows red. In addition to the visible red light, it's also giving off infrared light, which is what feels hot. When this strikes an object, say the sausages in the grill pan, it causes their constituent molecules to vibrate, raising their temperature.

Scientists' understanding of radiating heat had improved in the 1860s thanks to James Clerk Maxwell, who published a set of mathematical equations describing "electromagnetism."

For a sense of Maxwell's reasoning, imagine holding one end of a very long rope. It's stretched fairly tight and the other end is, say, a mile away. Jerk the end you're holding up and down. You see a kink travel away from you down the rope. Now move the end of the rope up and down continuously. A continuous undulating wave travels down the rope.

To see why, imagine the rope as a chain of tiny beads. Each is connected to the next by a short stretch of elastic. When you move the first bead in the chain, it pulls the one adjacent to it. That then pulls the one beyond it and so on. The up-and-down movement of the first bead is thus passed sequentially down all the beads, which looks like a wave moving down the rope.

How fast does the wave travel down the rope? It depends on how heavy the beads are and on the tension in the connecting elastic. Making the beads heavier will slow it down because it takes more effort to move them. Increasing the tension will speed it up. Each bead can pull harder on the next if the elastic between them is tauter. Intuitively, if you shake the end of a heavy, slack rope, the wiggles travel down it slowly. In contrast, waves will race down a taut, light guitar string at over one thousand kilometers an hour.

In Maxwell's imagination, empty space is filled with taut "strings" of this kind. They emanate from many of the particles that make up all the "stuff" in the world around us. Take, for example, the tiny negatively charged electron, a constituent part of all atoms. Imagine just one electron motionless in empty space. Tight strings stretch out from all directions through even the vacuum. Known as *electric field lines*, they're invisible and incorporeal but if you put another charged particle, like a positively charged proton, in a field line, it feels pulled toward the electron just as a bead in the chain feels pulled.

Now imagine the electron starts oscillating up and down. Just as the wave traveled down the rope, waves travel away from the electron down the electric field lines emanating from it.

So how fast do these electric field waves move? In one of the great insights of science, Maxwell identified how to estimate this. Take one field line stretching out from the electron. Imagine along its length, there are tiny compass needles. As the wave moves up and down along the field line, the compass needles swivel back and forth, toward it and then away from it. Readers may know an electric current flowing down a wire can have a similar effect, creating what's known as a magnetic field around it. Maxwell was saying that as waves move down electric field lines, they generate waves in an accompanying magnetic field. He pictured these waves at right angles to one another. For example, say the electric field wave oscillates up and down as it moves past you from left to right. Then the accompanying magnetic field wave will oscillate toward you and away from you. And, important, creating this magnetic wave takes effort just as moving the weighted beads in the rope took effort.

Maxwell's reasoning was intuitive, a hunch. But it had an enormous benefit. Remember with the wiggling chain, we could predict the speed at which a wave will travel along it by weighing one of its beads and by measuring the tension in the interconnecting elastic bands. Similarly, Maxwell could easily obtain measurements for their equivalents in field lines. The tension could be obtained by measuring how strongly two charged objects attract each other. The equivalent of the weight of a bead came from measuring the strength of the magnetic field created as a known current flowed down a wire.

Using these measurements, Maxwell estimated that these "electromagnetic" waves travel at about 300,000 kilometers per second. Lo and behold, that was remarkably close to measured estimates of the speed of light— too close to be a coincidence. It seemed highly unlikely that light "just happens" to move at the same speed as an electromagnetic wave; it seemed far more likely that light actually is an electromagnetic wave.

The point is any oscillating electric charge will emit an electromagnetic wave. Daylight thus exists because electrons in the sun are constantly being vibrated. They send waves down the field lines emanating from them. When these reach our eyes, they shake charged particles in our retinas. (This is otherwise known as *seeing*.)

Maxwell showed that the color of light is determined by the rate or the frequency at which the electromagnetic waves oscillate. The faster it does so, the bluer the light. Red light, the lowest-frequency visible light, is an electromagnetic wave oscillating 450 trillion times a second. Green light oscillates at a higher frequency, at around 550 trillion times a second, and blue light at around 650 trillion times a second.

Not only did Maxwell's theory describe visible colors, but it also predicted the existence of invisible electromagnetic waves. Sure enough, these were found from the 1870s onward. Radio waves, for instance, have frequencies that range from fewer than a hundred oscillations per second to up to around three million. The term *microwave* covers a range from there up to three hundred billion. Infrared sits between microwaves and visible light. When frequencies are greater than that of blue light, they are ultraviolet rays. Then comes X-rays, and oscillating up and down over a hundred billion billion times per second are gamma rays. The entire range, from radio waves to gamma rays, is called the electromagnetic spectrum.

Maxwell's discovery meant physicists knew in principle how the filament in a light bulb was made to glow. An electric current makes the filament hot. This in turn causes its constituent electrons to oscillate and emit electromagnetic waves. In fact, all objects emit some electromagnetic waves. Atoms are in constant motion, which means so are their electrons. For instance, at a healthy temperature of around 97°F, human bodies emit detectable infrared waves. Snakes, such as vipers, pythons, and boas, have evolved organs to detect such radiation to help them hunt and find cool places to rest.

The puzzle in the late nineteenth century was—what is the precise relationship between the temperature of an object and the frequencies of electromagnetic waves it produces?

For a sense of how physicists tackled this question, think of a potter's kiln. Like most objects, heating it causes electrons in its walls to vibrate. But it is helpful to study because the color inside the kiln can be easily correlated with its temperature. Dark red indicates that it's becoming quite hot. As the hue alters to orange, then to a yellowy-white, the temperature inside the kiln becomes higher. Intuitively, most of us would agree that "white hot" is hotter than "red hot."

What's happening is that at lower temperatures the kiln gives off only invisible infrared radiation. The kiln feels warm but doesn't glow. As the temperature rises, some higher frequency visible red light is emitted as well. As the kiln becomes hotter still, rising well above 1,000°C, out come the higher frequency colors, first some green and then a minute amount of blue. But because red light is still being emitted, what we see at very hot temperatures is a mix of red, green, and blue light, which our eyes interpret as orange, yellow, and yellowy-white, depending on the proportions of each color.

But even at very high temperatures, a typical kiln still gives off mainly infrared radiation. A tiny fraction of the electromagnetic energy it creates emerges as visible light. Almost none comes out as ultraviolet or anything beyond that. Also, whatever its temperature, a kiln emits very little energy at the lower microwave and radio frequencies.

To see what happens at higher temperatures, consider the light from the sun. This is rather like a giant kiln operating at over 5,000°C. At this temperature, the type of electromagnetic radiation being emitted does change. The sun emits some infrared, but most of the energy it produces appears at the higher frequencies of visible light.

This is the reason our eyes, and those of most animals, have evolved to be sensitive to red, green, and blue—these colors constitute the most abundant form of electromagnetic energy reaching us from the sun. Relatively small amounts of solar energy reach us at higher or lower frequencies so there would have been little evolutionary advantage in being able to detect those.

What happens at even higher temperatures, say at the 12,000°C found on the supergiant star Rigel? It emits over half its electromagnetic energy in the ultraviolet range. But even such a hot star emits relatively low amounts of super-high-frequency X-rays.

So, what lies behind this correlation between the temperature of kiln like objects and the frequency of electromagnetic radiation they emit? Answering this required Boltzmann's statistical ideas, and when the answer did come, it kicked off a series of events that transformed physics.

Max Planck, the soon-to-be catalyst of this transformation, had entered physics harboring no ambition to revolutionize it. Intellectually, he was attracted to absolute laws such as the first law of thermodynamics, which unambiguously states energy is always conserved. He disliked Boltzmann's probabilistic explanation for the second law. The increase of entropy shouldn't occur, Planck felt, simply because that was the most statistically likely outcome.

Planck felt that the behavior of radiant heat would provide a new understanding of the second law. Heat flow by convection or conduction was readily explained by the random movement and collisions of discrete particles. Radiant heat in the form of continuous waves of electromagnetic energy seemed different. He hoped it would explain the flow of heat without relying on the laws of chance.

To that end, Planck pondered the way devices like kilns create electromagnetic waves as heat causes electrons in their walls to oscillate. In the last few years of the nineteenth century, he worked hard to derive a mathematical equation that matched the observed relationship between the temperature of kiln-like objects and the frequencies at which they emit electromagnetic waves.

But then came an unexpected development. For in 1900, the Berlin authorities began to ask whether electricity or gas would be more economic for street lighting. Both use heat to create light, but which would be cheaper to run? The organization tasked with answering this, the Imperial Physical Technical Institute, was funded by the German government and housed in the city on land donated by the industrialist Werner von Siemens. In 1900, its scientists developed a device called a cavity radiator.

The cavity radiator was in essence a kiln the shape of a cylinder that is an inch and a half across and about fifteen inches long. It enabled extremely precise measurements of the intensity of light at different frequencies over a wide range of temperatures.

One of the scientists carrying out experiments at the Imperial Physical Technical Institute with these devices was a friend of Planck's named Heinrich Rubens. On the afternoon of Sunday, October 7, 1900, he visited Planck at home. Rubens had good and bad news.

On the one hand, in the visible light and shorter ultraviolet ranges, Planck's mathematics seemed to work. His equations accurately predicted how much of this higher frequency radiation emerged as the cavity radiator became hotter. The bad news was that at longer wavelengths they didn't work so well. At any given temperature, Planck's equations predicted *less* infrared light than was measured.

Rubens also reported another development: the English physicist Lord Rayleigh had come up with an explanation for the low-frequency end of the graph. Rayleigh had asked himself what size waves can fit inside a device such as a cavity radiator? Essentially, he argued that there was less room for long wavelengths than short ones.

Picture a taut guitar string. Pluck it exactly in the middle, and its lowest—fundamental—note will emerge. Pluck it near one end and the sound is different because, as well as the low note, higher harmonic notes emerge. This

Examples of guitar string "modes"

is because the string can oscillate simultaneously in different "modes." In the lowest mode, the middle of the string goes up and down. In the next mode, the string vibrates in an S shape. In the mode above that, a double-S shape, and so on. These modes are known as standing waves.

Electromagnetic waves similarly form standing waves inside cavity radiators. Remember, the device is a cylinder. Its ends are like the two ends of the guitar. Different modes will fit along the length of the cylinder just as the guitar string's modes fit along the instrument's length. But, argued Rayleigh, the size of the cavity radiator will discriminate against longer wavelengths.

Why? Because many more short wavelength modes can be fitted into the cavity radiator than long wavelength ones. Say the device is 60 cm long. The longest wavelength that will fit is 120 cm—that's the first mode with a peak in the middle of the device. The next-longest wavelength is the 60 cm one—that's the second mode and has two peaks. The third longest is 40 cm, the fourth longest 30 cm. So in the range 30 cm to 120 cm, only four wavelengths will fit. Now calculate how many modes whose wavelengths lie in the range 0.5 to 1.5 cm fit. The answer: 79 different wavelengths.

By this reasoning, Rayleigh concluded that the amount of long-wavelength radiation being emitted by the cavity radiator had to be less than that of shortwave radiation. His argument flowed naturally from the wave nature of light, and more important, it produced mathematical predictions that matched the data at low frequencies.

At high frequencies, however, it was wrong. Because there was theoretically no limit to how many short-wavelength modes could fit inside the cavity radiator, Rayleigh's method implied that even at low temperatures it should be full of ultraviolet light and X-rays. In reality, even at the hottest temperatures, such radiation barely existed.

So what does this mean? In a nutshell, Planck's mathematics were inaccurate for the low-frequency energy observed in the cavity radiator but were accurate for the high frequencies.

Rayleigh's analysis was the other way around. His mathematics were right for the low frequencies but hugely overestimated the high frequencies.

Frustrated by his inability to explain this discrepancy, Planck carried out what he later called "an act of despair. . . . I was ready to sacrifice any of my previous convictions about physics."

The upshot? After five years of labor, far from removing statistics from thermodynamics as he'd hoped, Planck was forced to extend their remit.

Ludwig Boltzmann had used statistics to explain how heat disperses as atoms and molecules collide. Planck found that only by applying the same statistics to oscillating electrons in the cavity resonator's walls could he derive an equation that accurately matched what was observed. In his historic 1900 paper, Planck acknowledged that he had to introduce "probability considerations into the electromagnetic theory of radiation, the importance of which for the second law of thermodynamics was originally discovered by Mr. L. Boltzmann." Planck's five-year project to prove Boltzmann wrong had failed.

In addition to using statistics, Planck had to make a strange assumption about the physical world. Think of the inside of a cavity resonator as a cave whose walls are lined with bells, each of which can ring at a specific pitch—from a low-pitched peal through to a high "ting." If a powerful earthquake shook the cave, all the bells would ring at roughly equal volumes.

In the cavity resonator, similarly, there are oscillators, typically vibrating electrons, which emit a wide range of electromagnetic radiation, from low-frequency radio waves all the way up to high-frequency X-rays. As the resonator's temperature rises, the heat makes the oscillators shake as the bells in the cave do. But now comes a key difference. Planck had to assume that it took far more shaking energy to make a high-frequency oscillator emit radiation than it took to make a low-frequency one do so. It was as if a high-pitched bell needed to be shaken considerably harder than a low-pitched one before it would ring out. If an earthquake struck a cave with bells like these, then the ringing of the lower and mid-range pitched bells would swamp the sound of the high-pitched ones.

In general, Planck reasoned that it requires large chunks of energy to make oscillators emit high-frequency radiation. Much smaller chunks are needed to make the low-frequency kind. Imagine two oscillators—one can emit infrared light, which has a frequency of 300 trillion cycles per second. The second can emit blue light, which has a frequency of 600 trillion cycles per second. The second will need to receive twice as much energy as the first before it will emit any light. The other consequence of this is that light from the oscillators comes out in discrete lumps. In the above example, this means that the smallest possible lump of the blue light contains twice as much energy as the smallest possible lump of the infra-red light.

Planck then combined these ideas with Boltzmann's style of statistics to explain how hot objects like cavity resonators, kilns, and even stars radiate electromagnetic energy. For a sense of how he did this, consider the following thought experiment:

Imagine a shop in which blue sweets cost $5 each, green sweets cost $3 each, and red sweets $1 each. Also available are some large but cheap colorless sweets that cost only twenty cents. Relatively few of the latter are stocked, as they take up so much room.

Now imagine that the shop's clientele has an average of $2 each to spend. Some have $3 and a few, $5. After a while, if you tracked the different types of sweets that had been bought, you would count relatively few blue, green, and colorless ones and a large number of red ones. If the clientele was richer, say, more have $3 and $5, then you would count more green and blue sweets being sold. The key idea here is that the relationship between the clientele's wealth and the color of the sweets being bought is not precise—it's statistical.

In the same way, the relationship between the heat energy in a cavity radiator and the light it emits is also inherently statistical. This analysis also explains the different nature of the electromagnetic radiation emitted by a super-hot star like Rigel and our sun. The former has more money in the form of energy. So, it's more likely to "buy" ultraviolet light. The latter is "poorer" and so can mainly afford visible light.

This discovery, way back in 1900, is heralded as the birth of quantum physics. The reason—Planck soon started referring to the chunks of energy absorbed and emitted by oscillating electrons as *quanta*.

Important, however, is that Planck thought of his quanta as being an artifact of the way vibrating electrons give off electromagnetic energy. He remained convinced that light was a continuous wave of electromagnetic energy whose intensity could take any value at any frequency. Another twenty years of intense scientific work would be needed before the idea that quanta are a fundamental property of nature took hold.

With hindsight, Planck's 1900 paper does mark a point of no return. The discovery of quanta in radiant heat could not be ignored. That's why the pre-Planck era is referred to as classical physics and the post-Planck as modern physics. Fair enough. But Boltzmann's role in this story is underappreciated. He's lumped together with pre-quantum scientists without acknowledgment of how crucial his work was to the quantum revolution.

In 1920, when Planck gave his Nobel Prize speech—the award was for his discovery of energy quanta in 1900—he said, "It gave me particular satisfaction, in compensation for the many disappointments I had encountered, to learn from Ludwig Boltzmann of his interest and entire acquiescence in my new line of reasoning."

The quote is deliberately misleading. Far from Boltzmann acquiescing with Planck's reasoning, Planck had acquiesced with Boltzmann's.

# CHAPTER FOURTEEN

# Sugar and Pollen

> Boltzmann is magnificent. . . . He is a masterly expounder. I am
> firmly convinced that the principles of the theory are right.
>
> —Albert Einstein

In 1905, one year before Ludwig Boltzmann's suicide, a scientific paper
was published that not only vindicated his ideas but established them as
part of the scientific consensus. That paper persuaded skeptics that atoms
and molecules are real, and that statistical analysis of their behavior can
explain the second law of thermodynamics. Unfortunately, it was published two or three years too late to rescue Boltzmann from his soul-
destroying arguments with the phenomenologists.

The paper was by a young man, still in his twenties, who had come
across Boltzmann's ideas in the latter's 1898 monograph, *Lectures on Gas
Theory*, which opened with the line "I am conscious of being only an individual struggling weakly against the stream of time." The young reader
of the book saw immense power in Boltzmann's arguments and wrote to
his fiancée, also a physics student, "Boltzmann is magnificent. . . . He is a
masterly expounder. I am firmly convinced that the principles of the theory are right . . . that the question is really about the movement of atoms
according to certain conditions."

The young man was Albert Einstein.

In the popular imagination, Einstein's science springs out of nowhere.
Aged twenty-six, the legend goes, he had his "miracle year" while working
as a clerk in the Swiss Patent Office in Bern. That was the year he came up
with his $E = mc^2$ formula and transformed physics. But what most forget
are the other equally fundamental papers that Einstein wrote during that

same miraculous year. Two were inspired by Boltzmann, and one helped persuade the scientific community that the Austrian had been right to believe in the existence of atoms and molecules.

Einstein had read Boltzmann's critics but wasn't deterred. Nor was Einstein deterred by his dire financial straits. In 1900, on graduating from the same Zurich university where Clausius had taught, Einstein hoped to be appointed assistant professor. But he hadn't performed well enough in his final-year examinations to obtain such a post. He was also falling head over heels in love with a fellow physics student, Mileva Maric, and as far as his professors were concerned, he had strayed by reading works by James Clerk Maxwell and Ludwig Boltzmann, whose ideas were considered either too advanced or too speculative for an undergraduate physics course.

Keen to set up a home with Maric, Einstein wrote numerous pleas to professors throughout Europe. Few bothered to reply, and those who did rejected him. "I will soon have graced every physicist from the North Sea to the southern tip of Italy with my offer," Einstein complained in a letter to Maric. To make ends meet he worked as a private tutor and as a temporary schoolteacher. Finally, a friend, Marcel Grossmann, advised Einstein that a job as "technical assistant level III" at the Swiss Patent Office in Bern had become available. Moreover, Grossmann's father knew the Patent Office director and penned a letter of recommendation for Einstein. The job wasn't glamorous—it entailed assessing patent applications on behalf of more senior staff. For Einstein, though, it promised a much-needed regular paycheck, and the prospect thrilled him. He wrote to Maric, "I'll be mad with joy if something should come of that."

Then Maric fell pregnant. The child, a girl, was born in early 1902, but her fate is something of a mystery as Einstein and Maric never told family and friends about the baby. Historians surmise that the couple gave up the child for adoption because the stigma of being unwed parents would have angered Einstein's parents and hindered his job prospects in conservative Switzerland. If the latter was partly the motivation, it succeeded, because in the summer of 1902, Einstein did obtain the job at the Swiss Patent Office in Bern. Maric joined him there in early 1903, and the couple married in January of that year.

Work at the Swiss Patent Office was so undemanding that, as Einstein wrote, "I was able to do a full day's work in only two or three hours." This left ample time for other activities, and by 1904, Einstein's life was hectic.

In addition to his job and private scientific research, Maric had given birth to a son, Hans Albert. The circumstance, being happier, meant both parents were thrilled.

Throughout this period from 1903 to 1904, Einstein remained convinced that he would make fundamentally important contributions to his chosen field of endeavor. To this end, he sent his first papers for publication in the *Annals of Physics*, the most prestigious scientific journal in the German-speaking world. These early papers demonstrate Einstein's fascination with thermodynamics. They are, in essence, generalizations of the statistical ideas that Boltzmann had spent so much of his life developing. With this work Einstein had turned himself into one of the world's leading experts on thermodynamics and laid the foundations of his "miracle year" of 1905.

The first paper of that year, which Einstein submitted on March 17, was entitled "On a Heuristic Point of View Concerning the Production and Transformation of Light." It's the paper Einstein regarded as revolutionary at the time, and the one that was cited by the Nobel Prize Committee when he received their award in 1921.

In one sense, this paper is a reaffirmation of Max Planck's 1900 paper applying Boltzmann's statistical analysis of gas molecules to the way objects glow when they are heated. To reiterate, Planck had concluded that when hot objects give off or absorb light, their constituent molecules only do so in tiny lumps. Planck, however, didn't think that was how light always behaves. In general, he believed it was a continuous flow of electromagnetic waves.

Einstein's paper, however, goes much further than Planck's, who by his own account had reached for Boltzmann's ideas as "an act of desperation." In contrast, Einstein embraces them. He turns them into what he calls a "heuristic" argument that light always exists as a stream of discrete particles. *Heuristic* means *discovered*. Einstein's saying that though no data definitively proves it to be so, it's very useful to imagine "that the energy of light is discontinuously distributed in space." Einstein arrives at this conclusion by taking Boltzmann's statistical definition of entropy further than Planck dared. Many aspects of light can be understood, according to Einstein, if its behavior is "interpreted in accordance with the principle introduced into physics by Herr Boltzmann, namely that the entropy of a system is a function of the probability of its state."

In his paper Einstein reminds his readers of Boltzmann's statistical explanation of the entropy of a gas. By assuming that a gas consists of tiny lumps, molecules, that are in constant motion, Boltzmann had showed that its entropy will increase by chance alone. Einstein uses these arguments to show that the way the entropy of light changes can also be explained if it is considered, like a gas, to consist of discrete particles. Just as the air in the room you're sitting in is made up of tiny particles, so, too, is the light that illuminates it. Having established the particulate nature of light with arguments that mirror Boltzmann's statistical analysis of a gas, Einstein concludes his paper by showing how this idea can render hitherto unexplained optical behavior "readily understood."

An example Einstein gives is the so-called photoelectric effect—when a light beam (or any electromagnetic radiation) strikes certain substances, it creates an electric current. The relationship between the frequency of the light and the amount of electric current generated was puzzling. In many cases, bright red light produces no current, whereas dim blue light, which oscillates at a higher frequency, does produce some. Faint ultraviolet light of an even higher frequency produces even more. Einstein explained this as follows: Light consists of lumps of energy, but the amount of energy in each lump depends on the frequency of the light. So, a lump of red light is smaller (i.e., contains less energy) than a lump of blue light. A lump of blue light is smaller than a lump of ultraviolent light. Shining red light on a substance, therefore, is akin to bombarding it with feathers. If a hundred feathers blew into your face, you could brush them off with little difficulty. Shining ultraviolet on something is equivalent to firing bullets at it. A single bullet will do far more damage than a hundred feathers. In the same way, a few particles of ultraviolet light will generate far more electric current than a large number of red-light particles.

Einstein's 1905 paper is regarded as one of the founding texts of quantum physics. But it would take another two decades before its ideas were fully accepted and understood by the scientific community. Only in the late 1920s was the word *photon* adopted to describe light particles. And light's wavelike behavior did not go away. Photons exhibit both particle-like and wavelike behavior—hence the phrase *wave-particle duality* and the many mysteries of quantum physics.

It is worth noting that Einstein mentions Boltzmann's name six times in his paper, including an entire section entitled "Interpretation of the

Expression for the Volume Dependence of the Entropy of Monochromatic Radiation in Accordance with Boltzmann's Principle." Yet the book from which Einstein learned this principle, *Lectures on Gas Theory*, is the one in which Boltzmann voices his fears that his ideas will be forgotten. Saddest of all: Boltzmann was still alive in 1905 but there's no evidence he was aware of Einstein's work. He died unaware of the epic consequences of his work.

When Einstein sent his light-quanta paper off to the *Annals of Physics*, he knew how revolutionary it was. However, he had yet to establish his academic credentials and keenly felt his lack of a doctorate. So, a few weeks after finishing his paper on light quanta, he completed his dissertation in his spare time and dispatched it to the University of Zurich. It, too, came from his investigations into thermodynamics, and its aim was to find evidence in support of Boltzmann's belief in the reality of atoms and molecules.

Einstein sought his evidence from a phenomenon that is difficult to explain if atoms don't exist. What piqued his interest is the way that water becomes more viscous (i.e., stickier) when sugar is dissolved in it. To feel this effect, move your finger through a bowl of sugar-water solution. It will take more effort than moving your finger through a bowl of plain water. This increase in stickiness, Einstein argued, can be explained if sugar and water are made of discreet particles. Even better, Einstein believed that by measuring the stickiness of sugar water and comparing it to that of pure water, you could estimate the size of the sugar molecule. To suggest that something as banal as sugar water hints at a deep underlying truth about the nature of reality was a bold claim.

In Einstein's imagination, sugar water isn't an unvarying, indivisible liquid. Rather, it is a mass of tiny water molecules, which he visualized as tiny spheres bumping into and jostling with one another. Interspersed within the water molecules are the larger spheres of the sugar molecules, which prevent the smaller spheres from moving freely. In effect, the water molecules are slowed down as they collide into the large sugar molecules. The consequence is that sugar water is considerably stickier than pure water.

Einstein then went on to show how this description made good quantitative predictions. His thesis is packed with mathematical equations, which guide the reader to an extraordinary conclusion—that two simple measurements of the behavior of sugar water allow one to estimate the diameter of an individual sugar molecule.

The two measurements are:

1. Compare the viscosity of a sugar solution containing a known amount of sugar with that of pure water.

2. Measure the "osmotic" pressure of the sugar solution. Picture a jar divided down the middle by a thin membrane. Dilute sugar water is on the left, concentrated sugar water on the right. Water will tend to flow through the membrane from left to right to equalize this difference. The pressure it exerts as it does so is the "osmotic" pressure.

These measurements are simple enough to do, and Einstein looked up typical values gathered by other scientists. Using their numbers, he estimated the diameter of a sugar molecule to be around one ten-millionth of a centimeter ($9.9 \times 10^{-8}$ cm). That's pretty close to modern measurements and the examiners at the University of Zurich were sufficiently impressed to award Einstein his PhD.

This paper alone, however, wouldn't be enough for Einstein to achieve his real goal of unambiguously demonstrating the reality of atoms and molecules. The problem was that though Einstein had estimated the size of sugar molecules, even on the best microscopes they were still too small to see, making it impossible to confirm his estimate. He still needed to find some aspect of the natural world where the atomic and molecular structure of matter really showed. So shortly after completing the sugar-water paper, Einstein began work on the second of his miracle-year papers. It was this work that most directly vindicated Boltzmann's belief that atoms and molecules are real.

Einstein again found what he was looking for in a seemingly mundane phenomenon. This time he settled on Brownian motion, as the puzzling behavior of tiny dustlike particles in water was known. It had been studied in the 1820s by a friend of Charles Darwin's, a botanist named Robert Brown, who had extracted tiny particles from cavities inside pollen grains and then mixed them with water. The particles didn't dissolve like sugar; rather they became distributed throughout the water in what looks like fine mist. Scientists refer to this as a colloidal suspension. These particles were tiny, only ten-thousandths of a centimeter across, but they were just big enough for Brown to observe with a microscope. When he did so, he noticed something strange—they wobbled and drifted slowly in the water. At first Brown thought this might mean the particles were alive. But he then rejected that idea on observing the same behavior when either fine sand or gold dust was suspended in water.

For most of the nineteenth century, Brownian motion remained an unexplained mystery, but few scientists considered it to be of fundamental importance. Einstein disagreed with that assessment, and in May 1905 he sent a paper to the German physics journal *Annals of Physics* showing that the only reasonable explanation for Brownian motion was that atoms and molecules are real. He submitted it merely two weeks after he had completed his PhD thesis on sugar water, implying he had been thinking about both problems at the same time.

To see how Einstein's argument unfolds, again picture water at the microscopic level. Everywhere there are water molecules, which look like tiny spheres. Every now and then you encounter a giant sphere, thousands of times larger than the water molecules. That's a pollen particle. What you also observe is frenetic activity. The water molecules are vibrating and wobbling, bumping into one another like bumper cars. The ones next to the pollen particle collide into it. It's as if tiny Ping-Pong balls are bumping into a giant beach ball from every possible direction

At first glance, it might appear that nothing interesting will happen. Because the water molecules' movements are completely random, and because they bump into the pollen particle from every possible direction, you might reasonably assume that all the collisions between the water molecules and the pollen particle cancel out, and that the pollen particle remains motionless. But Einstein argued this is not the case. Over short periods, by chance alone, more water molecules will collide into the pollen particle from one direction than any other. When that happens, the pollen particle will be nudged and will move a little. Then in the next short period, there will be a similar nudge but in a different direction. As this process repeats over and over, the pollen particle will drift in a series of random zigzag steps through the water precisely as Brown had observed back in the 1820s.

How, you might wonder, does this picture establish the reality of atoms and molecules? Here the originality of Einstein's thinking comes into play. He showed that with this molecular model of colliding spheres it is possible to predict how far a pollen particle will drift over a set time. And as pollen particles are big enough to see under a microscope, it should be possible to measure how far the real pollen particles drift. If the measurement were close to Einstein's prediction, that would be powerful evidence that the unseen water molecules are real objects.

Given the random nature of the collisions between the water molecules and the pollen particle, how did Einstein predict how far the pollen particles will drift? The answer comes from a specific kind of statistical technique known as the Drunkard's Walk. Imagine a drunk in a town square. Every now and then he takes a step in a random direction. The question is, Can you predict where the drunk will end up after a certain number of steps? It turns out that though you can't say in which overall direction the drunk will have traveled, you can predict how far he will end up from where he started. So, if our drunk started near a lamppost, you couldn't say whether he'd end up north, south, east, or west of it, but you can predict that after, say, an hour, he will be fifty yards from his starting point.

Einstein applied this logic to the pollen particles. They, too, take a step in a random direction every now and then as the water molecules barge into them. Using the Drunkard's Walk formula, Einstein showed that a pollen particle a thousandth of a millimeter across, drifting in water at 17°C, will move six-thousandths of a millimeter every ten seconds.

It's worth pausing to appreciate the magnitude of this statement. Einstein was saying that if atoms and molecules are real, it's possible to make a numerical prediction that can be confirmed or refuted by experiment. He'd found a way of testing a century-old debate about the reality of atoms with a simple measurement. Or, as he put it at the end of the paper:

"Let us hope that a researcher will soon succeed in solving the problem posed here, which is of such importance in the theory of heat!"

In just four years Einstein's prediction was verified by Jean Perrin, a gifted experimental physicist working in Paris. In 1909, Perrin published a paper describing meticulous experiments he'd carried out to measure how far particles drift in water. Perrin's attention to detail was impressive. He didn't use pollen particles because they are often of irregular shape, which means their diameter is hard to measure. (Einstein's prediction of how far the particle drifts depends on knowing the particle's diameter.) After much testing, Perrin instead settled on particles made from a resin known as gamboge, which is extracted from a tropical tree of the same name. (Gamboge is yellow and used to dye the robes of Buddhist monks.) Treating gamboge with methanol and water turns it into a fine powder

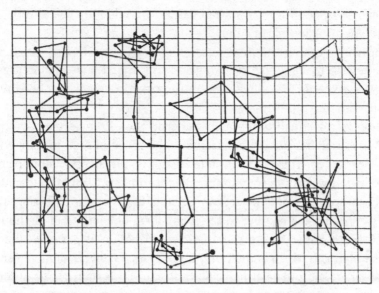

Brownian motion diagram from Jean Perrin's paper

made up of almost perfectly spherical particles whose diameter Perrin could accurately measure. They ranged from 0.5 millionths of a meter to 5-millionths of a meter across.

Perrin placed a suspension of gamboge particles in water under a microscope. To measure the distance the particles drifted, he projected the microscope's image onto a piece of graph paper, on which he then traced the particles' paths over a given time.

Perrin's measurements confirmed Einstein's prediction, within the bounds of experimental error. A decade later, all opposition to the reality of atoms and molecules among scientists evaporated. Wilhelm Ostwald, an enthusiastic energeticist and one of Boltzmann's most trenchant critics, relented, telling colleagues that Perrin's paper confirming Einstein's prediction had changed his mind. Even Ernst Mach seemed to have softened, for when Mach and Einstein met in 1912, an eyewitness later recalled that the two men agreed that the assumption that atoms and molecules are real could make useful and accurate predictions. But had Mach really changed his mind from the days when he had insisted, "I don't believe that atoms exist!"? It seems not, because after his death, Mach's son found a note in his father's papers: "In my old age I can accept relativity just as little as I can accept the existence of atoms."

Einstein's work on thermodynamics was not done. In that same year, 1905, he also made a historic extension to the first law—that energy is always conserved. This is reflected in the equation for which he is most famous: $E = mc^2$.

The $E$ in the equation represents energy, the $m$ represents mass, and the $c^2$ is a large but unvarying number—the speed of light multiplied by itself, or squared. This equation states that, though energy cannot be created or destroyed, it can sometimes take the unlikely form of solid matter. In other words, any solid object can be thought of as a form of congealed, highly concentrated energy; and that any form of energy can be thought of as a dilute mass. The most dramatic confirmation of this principle is a nuclear bomb, in which a small amount of mass is converted into an enormously energetic and destructive blast. Because $c^2$ is such a large number, a small amount of mass represents a vast amount of energy. In the bomb that destroyed Hiroshima, all it took was for about half a gram of mass, less than that of a paper clip, to turn into all that destructive energy.

Nuclear bombs were not on Einstein's mind when he derived $E = mc^2$. The equation is a logical consequence of an axiom he introduced into physics, namely that the speed of light is the same for all observers. If an observer on the ground shines a flashlight upward, she will measure the speed of light at a figure close to 300,000 kilometers per second. But what about a second observer traveling upward alongside the beam in a rocket almost as fast, at 299,000 kilometers per second? One might expect her to measure the beam's speed to be 1,000 kilometers per second. But she won't. She will also measure it at 300,000 kilometers per second.

Einstein came to this mind-bending conclusion from James Clerk Maxwell's description of light as an electromagnetic wave. Remember, light consists of two interleaved waves, one in an electric field and the other in a magnetic one, which are at right angles to each other. From laboratory measurements of how strongly electrical charges attract or repel and the strength of a magnetic field generated by an electric current, Maxwell was able to determine the speed of light.

To see then why this speed is the same for everyone, imagine two physicists, Alice and Bob, who wish to use Maxwell's reasoning to estimate the speed of light in a vacuum. Their laboratories are in space and both scientists wear spacesuits because there's no air in their labs. In the first lab, Alice makes the measurements of electric and magnetic field strengths that Maxwell described. From this she deduces a speed for light.

Bob performs the same measurements in his lab, which unbeknownst to him is moving past Alice's at 1,000 kilometers per hour. Einstein's argument was that Bob would obtain the same figure for the speed of light as Alice. Why? Because if he couldn't see out of his lab, he would have no idea he was moving relative to her. The vacuum feels the same no matter what speed you move through it. Because magnetic and electric fields exist in a vacuum, measurements of their respective strengths will have the same results irrespective of the relative speeds of the laboratory where they take place.

In 1905, Einstein couldn't definitively prove this. But he felt strongly that for the laws of physics to be worthy of that status, they should be consistent. It seemed wrong that two physicists would have to use different versions of Maxwell's theory of electromagnetism based on the relative speeds of their laboratories.

In 1905, in the third of his miracle-year papers, Einstein elucidated this idea, which has many extraordinary consequences. For example, time will flow differently depending on the speed at which you move. Observers moving at very different speeds will disagree on how far apart two objects are. Later in his fourth paper of that year, Einstein famously deduced that because the speed of light is the same for everyone, it also implies energy and mass can be turned into each other, in other words that $E = mc^2$.

Imagine again our two physicists, Alice and Bob. This time Alice is on the earth's surface and Bob is on a rocket. When a light beam is shot up from the earth into space, Bob takes off after it. The rocket engine is powerful, and Bob's speed steadily increases. But however fast he goes, he sees the light beam moving away from him at the same speed. He can never catch up with it. But what about Alice, back on the earth? It's important to note that for her, too, the speed of the light beam never changes. So how does she interpret that Bob never catches up with the light beam?

The answer: To Alice it looks as if the rate at which Bob's speed increases is slowing down. It appears to take him longer and longer to reach the speed of light. And, in fact, when Bob's speed is almost, but not quite, that of light, his speed no longer increases at all. His rocket engines are still firing as they did, but they seem no longer to be effective. How can this be?

To Alice it appears as if the energy from the rocket, *instead of increasing its speed, is increasing its mass.*

This thought experiment provides an intuitive explanation of why energy and mass must be interconvertible. Mass, like heat and movement, is another form of energy. So, whenever we cite the first law of thermodynamics and say that energy is always conserved, we must remember that all the mass in the universe is also a form of energy. Einstein himself saw his equation as an extension to the principle of energy conservation, first elucidated by James Joule and Hermann Helmholtz in the nineteenth century. "We might say that the principle of the conservation of energy," Einstein wrote in 1945, "having previously swallowed up that of the conservation of heat, now proceeded to swallow that of the conservation of mass—and holds the field alone."

Thanks to Einstein, the first law of thermodynamics had extended its rule. But why is it true? Why is energy conserved?

# CHAPTER FIFTEEN

# Symmetry

Gentlemen: I do not see that the sex of the candidate is an argument against her admission. . . . After all, the Senate is not a bathhouse.

—The mathematician David Hilbert

Before Noether's theorem, the principle of conservation of energy was shrouded in mystery.

—The mathematician and physicist Feza Gürsey

Emmy Noether was born in 1882 in Erlangen, a town in Bavaria in southern Germany. Three years younger than Einstein, she possessed a remarkable scientific mind and the determination to overcome the misogyny and anti-Semitism that dogged her career. Time and again Noether was denied posts that were her due, and she was eventually forced to flee Germany. She was brave and brilliant, and the majority of her male contemporaries didn't know what to make of her. In 1913, for instance, when she visited the mathematician Franz Mertens in Vienna, Mertens's grandson recalled, "Although a woman, [she] seemed to me like a Catholic chaplain from a rural parish—dressed in a black, almost ankle-length and rather nondescript coat, a man's hat on her short hair . . . and with a shoulder bag carried crosswise like those of the railway conductors of the imperial period, she was rather an odd figure." At the 1964 World's Fair in New York, a mural depicting the "Men of Modern Mathematics" showed eighty portraits—seventy-nine of men and one of Emmy Noether. Friends remembered her as fun loving, loud, and humorous and that she loved to dance. Colleagues recalled her generosity of spirit and her untrammeled passion for mathematics, the more abstract, the better. "This winter I'm giving a course on the hypercomplex, which is as much fun for me as it is for my students," she wrote to a friend. However, one colleague pointed out that she was "less well suited as instructor of large classes in elementary disciplines."

Fortuitously, Emmy Noether's father was professor of mathematics at the University of Erlangen, where he encouraged his daughter's talents. The university did not at first permit Noether to enroll officially as a student but as an "auditor." This meant she could attend lectures by professors who had granted permission for her to do so, but she could not be awarded an academic qualification. Then in 1904, in a step toward a more egalitarian educational policy, women were allowed to matriculate; Noether duly received her undergraduate degree and then her PhD. Her dissertation topic was an aspect of algebra known as the theory of invariants. Afterward, she moved to the University of Göttingen.

Göttingen, one of Germany's oldest university towns, had by the early twentieth century earned a deserved reputation for being one of the best places to study mathematics in Europe. Its head of the department, David Hilbert, one of the greatest mathematicians then alive, invited Noether to work there as a teacher and researcher on the strength of her PhD. In Hilbert, Noether found an ally who both appreciated her intellect and was willing to fight for her right to work in the face of opposition from the university's governing Senate, where members—particularly professors from the philosophy faculty—vehemently opposed officially recognizing Noether as an academic. Were she eventually to become a professor, they feared, she would be entitled to become a member of the Senate, which had never had a female member. Faced with this objection, Hilbert's response was scathing: "Gentlemen: I do not see that the sex of the candidate is an argument against her admission. . . . After all, the Senate is not a bathhouse." Hilbert then allowed Noether to lecture in his name, thus frustrating the Göttingen Senate's desire to prevent her from teaching. Noether worked in this way for four years without pay—her family covered her living expenses.

The year Noether arrived in Göttingen, 1915, was the same year Einstein published his general theory of relativity, overturning Isaac Newton's theory of gravity. Hilbert found the mathematical implications of

Einstein's work fascinating. In fact, one reason he had invited Noether to Göttingen was because she was an expert in the theory of invariants, a key technique Einstein had used in his work. Hilbert, in essence, asked Noether to assess the validity of Einstein's mathematics.

In this Noether succeeded, more than surpassing Hilbert's expectations. For while studying the mathematics of general relativity, she discovered why the first law of thermodynamics is true.

The theory of invariants is closely related to the idea of symmetry. Our world is filled with symmetries—the two halves of a human face are nearly mirror images of each other. Snowflakes look the same when they are rotated through an angle of sixty degrees. Many flowers exhibit symmetry, as does much art and architecture, including Leonardo da Vinci's *Vitruvian Man* and the Taj Mahal.

To mathematicians, symmetry is an example of "invariance." In other words, it's a process that, when applied to an object, leaves the object unchanged. It's easiest to use examples from geometry. A square, when rotated ninety degrees, looks the same as it did before being turned. A circle is completely symmetrical under rotation because, irrespective of the angle by which it is turned, it always looks the same. Note, however, that the idea of symmetry doesn't only apply to changes that take place in space; it includes changes that take place over time as well. So, if a circle looks the same as time passes, a mathematician will say it's "symmetric over time translation."

Thanks to her PhD research, Noether had become an expert in symmetry, and as she investigated Einstein's theories, she spotted a deep truth about our universe. Specifically, in a mathematical statement now known as Noether's theorem, she showed that for the laws of physics to be unvarying over time, energy must be conserved.

Take the simple example of a moving billiard ball colliding with a stationary ball. After the collision the two balls move apart. Their directions and speeds can be predicted by the laws of mechanics. If the two billiard balls collide at another time, whether on the next day or in two centuries, the same laws of mechanics will predict their subsequent behavior. It may seem a trivial or obvious point, but the takeaway is crucial: the equations of mechanics do not change over time. Noether proved mathematically that equations will only exhibit this symmetry if they are associated with a quantity whose value does not change. In other words, for time translation symmetry to exist in the laws of mechanics, something must be conserved. That something is what we call energy.

Noether's theorem goes far beyond energy conservation. It shows that whenever equations contain a symmetry, some quantity must be conserved. For example, the laws of mechanics do not regard one location in space as being more special than any other. Billiard balls follow these laws irrespective of where in the universe they are. This means the laws of mechanics have a spatial symmetry as well as a temporal one. For this to happen, a quantity called momentum is conserved. This is linked to the idea of inertia—the familiar feeling of being thrown forward when the vehicle you're in brakes suddenly. Put another way, this happens to ensure the laws of mechanics are the same everywhere in the universe. Other conserved quantities linked to symmetries include angular momentum and electrical charge.

An important aspect of Noether's theorem is that its converse is true; i.e., it implies that if the laws of mechanics are not time symmetric, then energy will not be conserved.

Evidence of this permeates the space around us. When the universe was about 379,000 years old, the first atoms formed, and all space was bathed in an orange-red light. Since then the universe has expanded a thousandfold, but that light still fills all of space. So why can't we see it? Because it has lost energy. Instead of the glow being orange-red, it now takes the form of invisible long-wavelength microwave radiation. This "cosmic microwave background radiation" was first detected by two Bell Laboratories scientists named Arno Penzias and Robert Wilson in 1964 and has been studied extensively since.

Think of the entire universe as a giant kiln, which, a long time ago, glowed orange-red. This light energy was most intense at a short wavelength of around a millionth of a meter, meaning its temperature was high, at around 2,700°C. Now this same light energy is considerably cooler—the wavelength of the glow is most intense at a long wavelength of 1.9 millimeters, which means its temperature has fallen to –270°C. Now when an actual kiln cools down, it loses heat to its surroundings. No heat is destroyed. But the universe has no surroundings to which it can lose heat. For it to cool down, heat disappears, which means energy is not being conserved. But this is what Noether's theorem predicts. In the early universe, the fabric of space and time were different than how they are now, and so the laws of mechanics, for example, were different than the way they are now. And that means in turn that energy is not conserved. In summary, Noether's theorem predicts that energy is conserved only when space and time remain unchanged.

When Einstein read Noether's work, he wrote to Hilbert, "I'm impressed that such things can be understood in such a general way. The old guard at Göttingen should take some lessons from Miss Noether. She seems to know her stuff." Since 1915, Noether's discovery has become a guiding principle for physicists. When the American physicist Richard Feynman gave his now-famous public lectures on science in 1963, he declared the link between symmetry and conserved quantities to be "a fact that most physicists still find somewhat staggering, a most profound and beautiful thing." Unfortunately, he didn't credit Noether for her discovery, wasting an opportunity to raises awareness of her work with the wider public. But we now know that much of the work underpinning modern particle physics derives from Noether's theorem.

At Göttingen, in the four years from 1915, as Noether developed her theorem, she continued to work and lecture unpaid under Hilbert's name. Finally, in 1919, the university authorities relented and employed Noether as a privatdozent, allowing her to teach officially and receive a salary. However, her interests now led her away from physics to more abstract ideas. For after her work on general relativity and symmetry, Noether's research focused on the foundations of mathematics. This work didn't have any direct relevance to physics, but it heavily influenced the subsequent direction of many aspects of mathematics, particularly algebra and topology.

Einstein, meanwhile, found himself being turned into the public face of science. While only physicists knew of Noether's theorem, the equation $E = mc^2$ became the scientific equivalent of "to be or not to be"—an iconic phrase that many quoted but few understood. Articles about the theory of relativity appeared in the *New York Times*. Max Fleischer, the creator of Betty Boop, produced an animation explaining how space could curve. Charlie Chaplin invited Einstein to Hollywood premieres.

Amid the hullabaloo of being a celebrity, Einstein's profound contributions to the theory of heat, thermodynamics, and their link to the new science of quantum physics were overshadowed. These topics, however, remained of great interest to Einstein himself. In the 1920s, working jointly with the Bengali physicist Satyendra Nath Bose, Einstein greatly extended the understanding of the statistical behavior of light "particles," which he had written about in his first *annus mirabilis* paper. Over the same period, the same topic became the crux of Einstein's long intellectual battle with the great Danish physicist Niels Bohr, over the implications of quantum mechanics. Though Einstein had helped to put down the foundations of the subject in his 1905 paper on light quanta, he grew wary of the direction in which younger physicists such as Bohr were taking it. The Dane was part of a group that included Werner Heisenberg, Wolfgang Pauli, and Max Born, who felt quantum mechanics represented a novel view of how nature worked at its most fundamental level. Einstein's reason for disagreeing is summed up in an oft-quoted remark he made in a letter to a colleague:

"Quantum theory yields much, but it hardly brings us close to the Old One's secrets. I, in any case, am convinced He does not play dice with the universe."

The quote refers to how quantum physics says that at the level of atoms, molecules, and light quanta, nature is inherently uncertain. For instance, you can never say precisely where an electron is. All you can say is that there is a certain probability that it is in a particular place. At first glance, this seems no different to the way Boltzmann and Einstein had thought about atoms and molecules. They accepted that one couldn't say with perfect precision how every single molecule of gas will behave at any given time. But, using statistical arguments, they could make a reliable prediction of what large collections of these molecules will do. But this view differs from the one developed by the so-called Copenhagen group—named for Bohr's home city. Boltzmann and Einstein had, in essence, used statistics to estimate what they could not measure. It is in practice impossible to know the position and speed of every molecule in a liter of air because there are too many to keep track of. In principle, however, if one had a sufficiently powerful microscope and a great deal of time, it would be possible. The Copenhagen group took a very different view. They argued that the behavior of objects such as atoms, molecules, and light quanta is inherently probabilistic. In other words, no matter how good your measuring apparatus, you will never be able to say exactly what these objects are doing. The best you can even hope for is a probabilistic estimate.

Einstein found the Copenhagen view deeply unsatisfactory. What's interesting is why he took this stance. Why did he find the probabilistic basis of quantum theory so unpalatable? The common view is that the probabilistic nature of quantum theory is at odds with his theories of relativity. Relativity may not be simple, but it is entirely deterministic. That means that if you know precisely how a system starts, you can, in principle, calculate where it will end up, in stark contrast to quantum physics, which says you can never know precisely either how a system starts or how it will end up. All you can say is something like there's a 50 percent chance it will start a certain way and a 50 percent chance it will end up a certain way.

But Einstein's early work on heat and the reality of atoms and molecules suggests his objection to quantum theory was not, in fact, due to its probabilistic nature. His papers on light quanta, sugar water, and Brownian motion embrace probability and statistics. Their central assumption is that one can't know for certain where a single molecule is or how fast it is moving, but that doesn't stop one from making reliable predictions for how a great mass of molecules behave.

And that's the point. Einstein didn't dislike probabilistic and statistical arguments. He'd cut his scientific teeth on them. But to him, statistics and probabilities were a way of grasping the underlying reality hidden from plain sight. Think of the pollen particles in Brownian motion. Einstein couldn't predict with 100 percent accuracy how they would behave, but he could make a reasonably good statistical estimate. Though the behavior of the pollen particle wasn't entirely predictable, it did give an insight into the invisible world of atoms and molecules that underpinned and drove that behavior. And this was his contention with Bohr's analysis of quantum theory. Bohr was saying that nature at the quantum level is probabilistic and that there is no deeper, underlying reality. Whereas Einstein, drawing on his experience of Brownian motion, would have felt that statistical behavior implied a deeper fundamental truth. This truth wasn't directly visible, just as atoms weren't; but for Einstein, not to even acknowledge the existence of this deeper layer of reality was an anathema.

Einstein continued to take an active interest in the science of heat for much of his career, in line with his belief that science should serve society. These days, Einstein is so routinely characterized as the absentminded professor that his practical, inventive side is little appreciated. After all, Einstein had been raised in a home where inventing, building, and tinkering with machines was commonplace. His father, Hermann, and Einstein's uncle Jakob had run a small electrical engineering firm that built dynamos and electricity meters, and though the business struggled and eventually closed, it left the young Albert with a lifelong interest in technological innovation.

Einstein's first partner in this was an inventor named Rudolf Goldschmidt, with whom Einstein registered a patent for electromagnetically driven loudspeakers in 1928. Then when a mutual friend, a singer called Olga Eisner, began suffering from deafness, the pair designed her a hearing aid.

Neither invention advanced beyond the design phase. The technology with which Einstein progressed furthest was directly inspired by his interest in heat and thermodynamics. During the late 1920s and early 1930s, he worked to help design, patent, and market a refrigerator. At the time, these devices were thermodynamically speaking quite advanced, but they used toxic chemicals such as ammonia, methyl chloride, or sulfur dioxide as their coolant fluid. Refrigerator pumps, if their seals leaked, would release these toxic chemicals into their owners' homes with devastating consequences. In 1926, Einstein had read a harrowing newspaper account of one Berlin family, including several children, who had died from the fumes emanating from their malfunctioning refrigerator. The story prompted Einstein to try to design a safer one.

As collaborator, Einstein chose a former student named Leo Szilard. Born in Budapest in 1898, Szilard's talents for physics and mathematics had showed early when he won a Hungarian national prize in mathematics at age eighteen. Soon after he began studying physics at the Friedrich Wilhelm University in Berlin, where Einstein lectured. A lifelong and consequential friendship began. Szilard's PhD, written in 1922, the first paper to detect a link between thermodynamics and information theory, was rated the best that year, and by the mid-1920s, Szilard and Einstein had become close friends. Both gifted scientists, the two men had similar values, which included a firm belief that science should serve society. So, when Einstein felt better refrigerators were needed to avoid unnecessary deaths, he called Szilard.

The pair made a promising start with a device that combined safety, simplicity, and cheapness and that secured backing from a company in Hamburg named Citogel. The name means *quick freeze* in Latin. A small inner chamber (2) sat in the middle of a larger cylinder (13) that contained

REICHSPATENTAMT

# PATENTSCHRIFT

№ 527 080

KLASSE **17a** GRUPPE 7

S. No. 88 / 17.a

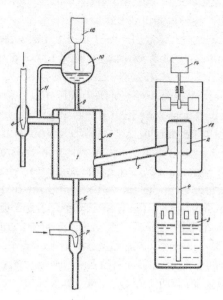

One of Einstein and Szilard's
refrigerator patents

whatever needed cooling, say ice cream. A chemical named methanol was then evaporated in the inner chamber, thus cooling the surrounding cylinder. A small pipe (5) then conveyed the now gaseous methanol to another cylinder that was connected to an ordinary water tap. From this water flowed, dissolving the methanol and flushing it away. The upside of this design was that it needed no power other than tap-water pressure and that methanol fumes aren't toxic in small quantities. The downside was that the methanol coolant was not reused. Once its cooling work was done, it was flushed away down the sink. Einstein and Szilard figured that the low cost of methanol meant this wouldn't put off potential customers.

Citogel named the device the Volks-Kuhlschrank, or the people's fridge, presenting it at the spring fair in Leipzig in March 1928. The company's share price shot up by 50 percent.

But the people's fridge did not catch on. First, methanol was more expensive than predicted in the business plan, and second, the design relied on constant pressure from domestic water taps. In 1920s Germany, this varied so much from building to building, even from floor to floor, that those testing the device complained of poor reliability. "The invention never did get on the market," reported Szilard.

Back at the drawing board, Einstein and Szilard came up with their most imaginative idea—a device that worked like a conventional refrigerator, but which had a revolutionary compressor design. Remember this key component warms gaseous coolant and then pumps it into a condenser so it can then release the heat it has absorbed from the refrigerator's cold interior to the outside air. Unlike a conventional compressor, which used spinning metal blades to work, the Einstein-Szilard device used a varying electromagnetic field, generated by an electric coil, to make liquid metal in a sealed cylinder move back and forth. This motion drove the compressor. The advantage from the safety perspective was that all the potentially dangerous substances—the refrigerator's coolant and the liquid metal—were permanently contained within stainless steel pipes and cylinders. There was no seal that might become damaged and leak.

Having begun his working life as a patent clerk in Bern, Einstein oversaw the intellectual-property aspect of the collaboration, registering forty-five patents in no less than six different countries for their various refrigerator designs. Their work also attracted attention, for in late 1928, Allgemeine Elektricitäts-Gesellschaft (German General Electric Company or AEG) funded engineers in their Berlin research laboratory to build a prototype Einstein-Szilard compressor. They also agreed to pay Szilard patent royalties and consulting fees amounting to $3,000 a year, which is around $40,000 today. Szilard deposited the money in a joint account he had opened with Einstein.

Einstein took a keen interest in the AEG prototype's progress. He made regular trips to the company's laboratory, and Albert Korodi, one of the engineers developing it, later recalled visiting the great scientist in his apartment over a dozen times to discuss their work. Like all prototypes, the compressor had teething problems, and early versions were exceedingly noisy. In the words of Szilard's friend Dennis Gabor, it "howled like a jackal." Another witness said it "wailed like a banshee." Korodi, the engineer, was rather more charitable and described its sound as more like "rushing water." The engineers worked out a way to reduce the noise, and on July 31, 1931, a fully functioning prototype went into operation in the AEG research institute in Berlin. It ran safely and continuously and was considered a success.

So why don't today's refrigerators use Einstein-Szilard compressors? The answer is that while AEG in Berlin was developing the device, General Motors invented a new coolant called Freon at their research laboratory in Dayton, Ohio. As Freon was considerably less toxic than previous coolants, switching to it was a much cheaper way for manufacturers to improve refrigerator safety than by mass-producing a radically new pump. Little did manufacturers know that Freon, part of a class of chemicals known as hydrofluorocarbons, would cause a hole in the ozone layer in the earth's upper atmosphere. Modern coolants, essentially chemical variations on Freon, don't.

The other reason that the pair stopped work was the increasing social turmoil in Germany. By 1930, the country's economy was in a woeful state and unemployment was soaring. Szilard sensed what was coming. In September of that year, he wrote to Einstein, "if my nose doesn't deceive me, from week to week I detect new symptoms that peaceful development in Europe is not to be expected in the next ten years. . . . I don't even know if we will finish building our fridge in Europe." Less than three years later, Hitler became chancellor. Einstein, who happened to be abroad at the time, announced he would not return to Germany. Meanwhile Szilard, who was also Jewish, escaped on the train from Berlin to Vienna. The following day, Nazi soldiers held back passengers on the same train whom they deemed "non-Aryan" and seized their most valuable possessions.

After taking power, the Nazi government immediately directed a boycott of Jewish businesses, encouraged violence against Jewish people, and enacted a regulation called the Law for the Restoration of the Professional Civil Service, which forbade any German government organization from employing Jewish workers. Thousands of German Jewish academics were fired. Among them was Emmy Noether, who had fought so hard to become an academic at Göttingen. In April 1933, she received a notice from the Prussian Ministry for Sciences, Art, and Public Education: "With reference to paragraph 3 of the statutes for professional civil servants of April 7, 1933, I herewith withdraw your permission to teach at the University of Göttingen."

Astonishingly, Noether's immediate response was to stay in the town and continue to teach, once again unofficially, at her home. One student had the gall to turn up in the brown-shirt uniform of the Nazi paramilitary group the Sturmabteilung, or the SA. Noether showed no outward sign of concern.

Fortunately, Noether's reputation as a mathematician had spread, and on hearing of her dismissal, Bryn Mawr College in America offered her a post. Noether then emigrated with financial assistance from the Rockefeller Foundation. From late 1933 until her death two years later from complications following surgery, she taught at Bryn Mawr and Princeton. A year after Noether left Göttingen, David Hilbert, now seventy-one, an invalid who had seen his closest friends and colleagues driven out of the country, was asked by the Nazi minister of education, Bernhard Rust, whether "the Mathematical Institute really suffered so much because of the departure of the Jews." "Suffered?" Hilbert replied, "It doesn't exist any longer, does it!"

Yet back in 1933, despite the gathering storm, there were a few glimmers of hope. Shortly after fleeing to Vienna, for instance, Leo Szilard happened to meet the then director of the London School of Economics, William Beveridge, who is better known now for founding Britain's postwar welfare state. Szilard and others, such as the economists Ignaz Jastrow and Jacob Marschak, urged Beveridge to address the plight of Jewish academics in Germany. As a result, the Academic Assistance Council was set up shortly afterward to rescue scholars and scientists who had lost their livelihoods.

At Beveridge's suggestion Szilard moved to London, where he worked tirelessly to promote the work of AAC and to mobilize wider international support for scholars persecuted by the Nazis. In London, Szilard supported himself on the money he'd earned from his work on the refrigerator with Einstein. Without these funds, Szilard could not have fled Germany, let alone devoted so much time to helping people escape the Nazis. In unimagined ways, Einstein and Szilard's fridges did save lives.

# CHAPTER SIXTEEN

# Information Is Physical

A decidedly unconventional type of youngster.
>—Scientist Vannevar Bush's description
>of Claude Shannon

Every time you search the internet, the oceans and atmosphere become a tiny bit warmer. The energy required to perform a hundred or so Google searches, a number I easily surpass in a working week, would heat the water needed to make a cup of tea. According to Google's own data, their electricity consumption in 2018 was just over 10 million megawatt hours, which is about the same as a small country like Lithuania. Data centers worldwide use about 1 percent of global electricity. Information and communication technologies contribute to more than 2 percent of the world's carbon emissions, which is roughly the same as that of the aviation industry. Some studies expect this sector to use 20 percent of the world's electricity by 2030.

Energy is needed to drive the machines that process and send our information, and that energy ends up as waste heat, which is dissipated into the planet's oceans and atmosphere. The information age, like the Industrial Revolution of the nineteenth century, is for the most part enabled by water and steam. These substances drive the turbines that generate much of the electricity that carries our information. The cooling systems that remove heat from the machines that process information in vast server farms also rely largely on the thermodynamic properties of water.

Recent years, therefore, have seen a surge of interest in the link between information and thermodynamics. Is the connection merely a practical one, a consequence of the way we chose to build computers and communications systems? Or is the connection more fundamental? If so, how can something as nebulous as information, which includes concepts as varied as thoughts, words, music, images, films, and even genes, be linked to well-defined scientific quantities such as energy and entropy? And will processing information always dissipate heat and increase the entropy of the universe?

The pioneers of the information age never considered such questions. Much as the first people who built steam engines and refrigerators didn't understand their science, so the engineers who created the earliest technologies with which we now communicate didn't understand their fundamental principles.

In the early 1910s, the American Telephone and Telegraph Company, or AT&T, was struggling. Ma Bell, as the company was known, had originally been founded by Alexander Graham Bell's father-in-law in 1877, but by the early 1900s, it could not compete with the many smaller rivals providing local calls in America's cities. The bullish reaction of AT&T's senior management was a plan to build the first long-distance telephone service, one that would connect America's coasts.

This was an enormous technological challenge. The signal created by a telephone's microphone is complex—necessarily so as it is an electrical copy or analog of the many different sounds that constitute speech. Our vocal cords create variations in air pressure, which a microphone tries to replicate as variations in an electric current. An earpiece turns the latter back into air-pressure variations—i.e., sound. There are many pitfalls. The microphone and earpiece in the telephone handset add distortion, as do the network's connecting cables. And even with the best-made wires, the signal's power decreases as it travels and soon becomes lost in background noise, as we know the random electrical effects that challenge any signal flowing down a wire. There was a hint here of the connection between information and thermodynamics and in particular the second law. Although engineers at the beginning of the twentieth century didn't realize it, the tendency for a message to become garbled and unintelligible is similar to the way heat tends to dissipate.

To meet the challenge of long-distance telephony, AT&T hired a team of scientists who had all been taught by Robert Millikan, America's leading physicist, who in 1923 would win the Nobel Prize. It proved an inspired move—the strategy paid off when Millikan's protégés perfected a device known as the thermionic valve (or the vacuum tube as it's known in America). This allowed telephone engineers to design amplifiers that boosted signal levels while keeping the background noise tolerably low. Conversations could now travel along three thousand miles of copper wire, held in place by 130,000 wooden poles, that spanned the American continent. On January 25, 1915, as a publicity stunt, the sixty-seven-year-old Alexander Graham Bell was asked to repeat the words he'd spoken thirty-nine years earlier in the world's first telephone conversation he'd had with his old assistant Thomas Watson. Then the pair had been in the same building. Now Bell was in New York, Watson in San Francisco. "Mr. Watson, come here, I want you," said Bell. "It would take me a week to get there now," Watson replied. Thermionic valves became the mainstay of telephony in the first half of the twentieth century, enabling global communication via cables or wirelessly with the use of radio waves.

There's an analogy here to the early days of steam. Remember how inefficient engines were then, wasting well over 90 percent of the heat they produced. Not knowing how to improve them, people burned more coal. The thinking behind thermionic valves was similar, for they didn't reduce noise levels so much as raise the strength of signals so they were no longer swamped. In both cases, engineers were pumping more energy into the system to overcome the waste of most of it.

Nonetheless, thermionic valves revolutionized communication. Their invention also convinced AT&T's senior management that it was worth funding a permanent research institute. Founded in 1925 with an annual budget of $12 million ($150 million in today's money), this was R&D on a scale that no other corporation had before attempted. In a vast thirteen-story yellow-brick building on West Street in Lower Manhattan, two thousand technical experts worked on product development; another three hundred worked on basic and applied research. Hired because their expertise spanned disciplines as varied as fundamental physics, chemistry, materials science, meteorology, and even psychology, the scientists were deliberately not given specific goals. Fundamental research carried out by an interdisciplinary team would, it was believed, result in scientific and technological breakthroughs.

It did. In spades. Within a few years, the lab's engineers had made huge breakthroughs in fax transmission, television broadcasting, and cryptography.

Among the many brilliant and eccentric scientists who worked at the labs was an unusual young man named Claude Elwood Shannon, who loved puzzles, tinkering with electrical equipment, and juggling. He saw the world with a clarity that few possess. His ideas would enable us to build the data networks that define our age and pin down for the first time what information means.

Claude Shannon was born on April 30, 1916, in the town of Gaylord in the middle of a plateau that dominates the northern half of the US state of Michigan. With a population of around three thousand, the town was "small enough," Shannon said later, that "if you walked a couple of blocks, you'd be in the countryside." The main sources of employment were the lumber industry or the potato fields. Claude's father, Claude senior, was an ex–traveling salesman from New Jersey who seemed to have a hand in every aspect of the town's life. Active in the Methodist Church, he was also a Freemason, a probate judge, and ran a business that both sold furniture and arranged funerals. Shannon's mother, Mabel, was intellectually gifted and determined. Despite growing up poor, she obtained a college degree at a time when it was hard for women to do so, and shortly after her children were born, Mabel became principal of Gaylord High School. Yet despite being a highly regarded principal, Mabel was fired in 1932 during the Great Depression. The school board decided that "when a husband was capable of making a living, it would be unfair competition to hire married women."

It was Claude junior's older sister, Catherine, who first showed an aptitude for mathematics, so excelling at school that she was set on a track that led to her becoming a professor of the subject. When Catherine entertained her younger brother with mathematical puzzles, he was hooked. He also began to build radios and remote-controlled boats, fix his neighbors' electrical equipment, and construct a private half-mile-long telegraph system that used barbed wire to carry the signal from his house to a friend's. By the time he graduated from high school, Shannon was already well versed in the practical aspects of telecommunications. He had seen for himself the pros and cons of a Morse code–based system such as telegraphy compared with those of analog telephony.

In 1936, after graduating from the University of Michigan with degrees in engineering and mathematics, Shannon enrolled as a master's student at the Massachusetts Institute of Technology. Barely twenty, he was a shy and skinny five feet ten, weighing under 140 pounds, and his awkward laughs were said to sound like coughs. He could solve cryptographic puzzles that baffled his fellow academics but needed to write out simple arithmetic problems longhand. A consensus formed in the corridors of MIT: Claude Shannon was special.

So special that when Shannon enrolled for flying lessons, the instructor, also an MIT professor, felt he shouldn't be permitted to take the course as the risk of losing such a remarkable brain in a crash wasn't worth taking. MIT's president overruled the flying instructor, and Shannon did obtain his pilot's license. More significant for his career, the dean of engineering at MIT, Vannevar Bush, one of America's most influential scientists, took it upon himself to nurture Shannon's talents. "A decidedly unconventional type of youngster," Bush wrote to a colleague. "He is a very shy and retiring sort of individual, exceedingly modest, and who would readily be thrown off track."

The first task Bush set Shannon at MIT was operating and developing the engineering department's experimental analog computer. Unheard of these days, analog computers were considered promising in the first half of the twentieth century. MIT's machine was so big that it occupied an entire room. It worked by sending continuously varying electrical currents through circuits based on thermionic valves. They were well suited to performing calculations related to a branch of mathematics known as calculus. Shannon fell in love with the machine, which he learned to program by pushing and turning an array of rods, pulleys, gears, and disks.

One evening, while standing minding his own business during a party, Shannon felt a handful of popcorn hit him in the face. He looked up to see a woman, who asked him why he wasn't among the revelers. He replied that being where he was, by his bedroom door, meant he could hear his favorite music playing on his gramophone. She asked him, "Bix Beiderbecke, you got him?" He replied, "My favorite."

A whirlwind romance followed. The woman, Norma Levor, herself only nineteen, a student at nearby Radcliffe College, had had an upbringing that was the polar opposite of Claude Shannon's. She'd grown up in a penthouse situated near New York's Central Park, the daughter of a pincushion heiress and an importer of fine Swiss fabrics. Something of a rebel, Levor had run away to Paris in the summer of 1939 and only returned home as her parents feared for her safety with war in Europe looming.

She was also actively involved in New York City's left-wing political scene. That night in the autumn of 1939, however, Levor later recalled, she fell for Shannon's "Christlike looks." Religious disbelief, however, was the one thing that this unlikely couple did share. When Levor told Shannon that she was an atheist, he replied, "How can you be anything else?"

The affair moved quickly. According to Levor, shortly after they met, Shannon took her one night to the room that housed the MIT analog computer, to which he had a key, and there, amid the cabinets of electrical relays and thermionic valves, they made love. Levor was attracted to the way Shannon loved poetry and had an artist's sensibilities despite being a scientist, a characteristic which informs his later work. "So sweet, so full of fun and such a joy to be with" was how Levor remembered Shannon then. He even took her flying thanks to his recently acquired license, "scaring me shitless." A few months later, in January of 1940, the couple married and honeymooned in New Hampshire, an event spoiled by an anti-Semitic hotelier who wouldn't let them have a room because Levor was Jewish.

Later that year, the couple moved to Princeton, where Shannon had been granted a fellowship at the Institute for Advanced Study. Here the pair rubbed shoulders with some of the world's greatest mathematicians and physicists, Hermann Weyl, John von Neumann, and Einstein, who had made Princeton his home after having been driven out of Germany in 1933. For the Shannons, however, life took an unhappy turn. Their affair, which had blossomed so quickly, self-destructed just as fast. To Levor, Shannon changed. He found the atmosphere at Princeton alienating and stifling and his natural joie de vivre evaporated. "I tried to get him to go to an analyst, but he wouldn't," Levor later recalled. "He just got grimmer and grimmer. I began to feel that unless he could do something about it, I couldn't stay with him." Levor is the only person who's written and spoken about this side of Shannon's character—he's normally described as a playful eccentric—but then she's the only person with the sensibilities of a writer who knew him well.

There were other pressures on the marriage. Levor found life as an academic's wife unfulfilling and yearned for a career in her own right as a screenwriter. She and Shannon divorced barely a year after they'd married, and Levor moved to Los Angeles. There, she became a screenplay writer and was remarried—this time to a left-wing writer who shared her political outlook. The couple joined the Communist Party and then spent the next thirty years as blacklisted American exiles in Europe.

Claude Shannon, devastated by Levor's departure, made a much shorter journey, from Princeton back to New York City. Here, he, too, found his life's destiny. For in the summer of 1941, he started work as a mathematician at Bell Labs.

Then on December 7 that same year, the Japanese air force bombed Pearl Harbor. With those at the Bell Labs applying their minds to the war effort, Shannon was assigned to a top-secret project—a system for encrypting wireless telephone calls called SIGSALY. The aim was to create a secure encrypted transatlantic communication channel for the highest echelons of the Allied forces. Shannon's role was to analyze its encryption method to ensure that it was robust.

The first SIGSALY call between London and Washington took place on July 15, 1943. The system was a resounding success. Over the course of the war, SIGSALY enabled about three thousand high-level telephone conversations, including some between Roosevelt and Churchill. Though German radio operators picked up these messages in their encrypted form, they didn't decipher a single word. After the war, Allied investigators discovered a memo from the German code-breaking department assessing the possibility of decrypting SIGSALY. It read, "There is not much to be gotten from them now."

Working on SIGSALY gave Shannon firsthand experience of cutting-edge communications technology. He responded to this as Carnot had to the steam engine, seeing past the engineering challenges to the fundamental ideas. What exactly is a message? he wondered. How long or short does it need to be to communicate an idea? Is there an unambiguous mathematical way of measuring the size of a piece of information? After spending the day working at Bell Labs, he would go home to his Greenwich Village apartment and work late into the night on his private calculations.

There was another consequence of working on SIGSALY. For in the later months of 1942 on into 1943, Shannon met, almost daily, the one other person in the world in the fields of cryptography, communication, and computing who was his equal and his intellectual soul mate, the great British mathematician and code breaker Alan Turing.

By late 1942, Alan Turing had already played a pivotal role in helping the British crack Enigma, the encryption system used by the Germans

to protect their military communications, thereby establishing his reputation as Britain's leading cryptographer. So, when the American intelligence authorities informed their British counterparts about SIGSALY,
they sent Turing to Bell Labs to vet it.

Shannon and Turing didn't work together on SIGSALY—their respective governments' obsessions with secrecy meant that the one topic the two
could not discuss was code breaking. Nonetheless, they quickly formed a
lasting friendship. During the winter of 1942–43, they met almost daily
for tea in the Bell Labs cafeteria. "Turing and I had an awful lot in common," Shannon later recalled in understated fashion. "We had dreams.
Turing and I used to talk about the possibility of simulating entirely the
human brain, could we really get a computer which would be the equivalent of the human brain or even a lot better?" The two men also discussed
the principles underlying all forms of communication. "I had talked to
him several times about my notions on information theory, I know," said
Shannon later, "and he was interested in those."

With the war's end, Shannon's intellect was set free. As was his heart, for
he started dating Betty Moore, who would become his second wife. A gifted
mathematician, Moore worked as a human computer, carrying out mathematical calculations for the engineers at Bell Labs. The pair shared a love
of jazz and tinkering with electronics. Soon Moore had become a sounding
board for Shannon, and she encouraged his eccentricities, including juggling
and haring around the corridors of Bell Labs on a unicycle. The company, for
their part, allowed Shannon his fun, giving him no targets and encouraging
him to be himself. "They never told me what to work on," he recalled.

The strategy was spectacularly successful. The environment at Bell
Labs enabled Shannon's remarkable brain to put together his varied
experiences—barbed-wire telegraphy, SIGSALY, his conversations with
Alan Turing—and come up with one of the greatest scientific insights of
the modern age. In 1948, he revealed his thoughts in a paper entitled "A
Mathematical Theory of Communication," published in the Bell Labs technical journal.

Less than thirty pages long, Shannon's paper enabled humans, for the
first time, to measure information in a completely objective and clearly
defined way. What does this mean? A photograph, a novel, and a painting
are examples of information. Shannon gave us a way to numerically compare their relative sizes. The importance of this idea is immense. It means

that we can, for example, quantify all the world's telephone calls and work out exactly what we must build in order to convey them. But the benefits go far beyond the practical. Shannon found an objective definition of information much as William Thomson had for temperature with his absolute scale back in Scotland in the 1850s.

The first striking thing about Shannon's paper is its approach, which mirrors the one taken by the pioneers of thermodynamics in the nineteenth century. Much like James Clerk Maxwell, Ludwig Boltzmann, and Josiah Willard Gibbs had, Shannon started with the principles of statistics. He showed that the same laws of chance that explain the flow of heat explain the flow of information.

The purpose of communication, Shannon began, is "reproducing at one point either exactly or approximately a message selected at another point." But then he takes a surprising tack, arguing that to quantify information we must ignore the meaning of the message. "These semantic aspects of communication are irrelevant to the engineering problem," he states. This seems cold and self-defeating at first, but is in fact liberating. It makes Shannon's theory universal: by divorcing meaning from message, he found a way of measuring the size of all possible messages. With heat, Thomson's temperature scale is useful because it's independent of the substance being measured. We can meaningfully say, for example, that an ingot of iron, a glass of water, and a cabbage are all at 300 kelvin. So, too, Shannon shows us how the informational size of a piece of text, a picture, and a genome can be evaluated.

Shannon's next crucial assumption is that all communication is encrypted. The only difference between two people using a system such as SIGSALY, say, and two people speaking in plain English is that in the former case only the pair who are speaking know how the message has been encoded, while in the latter case, the message is encoded in English sounds, the meanings of which are known to all speakers of the English language. This may seem an obvious point—all I'm saying is that before you can understand a language, you must first learn it—but it's important to Shannon's argument, which relies on the fact that for people to communicate, they must first agree on how they're encoding their messages.

Then comes a startling idea. Shannon argued that any and all messages can be communicated as the answers to a series of yes-and-no questions. There are no exceptions. Every piece of information can be communicated in an extended form of the game Twenty Questions, where player 1 thinks of a famous person and player 2 has to guess her or his identity by asking twenty questions to which player 1 can only answer yes or no.

Shannon showed that if you lift the limit of twenty and permit the player to carry on asking questions, she or he will always be able to work the answer out.

To see how this works, imagine that one person, Bob, wants to send the message *help* to someone called Alice using only yes/no questions. (These two names are commonly used by information scientists to typify the senders and receivers of information.) To make it realistic, let's say that the only way Bob can signal Alice is by turning a flashlight on and off.

Assume that Alice and Bob read English and agree on the way the alphabet is usually listed, from *a* to *z*. In front of each of them is the list:

*abcdefghijklmnopqrstuvwxyz*

Alice and Bob agree on the following rules. At a specified time, say 1:00 p.m., Bob will either flash his light on or leave it off. Exactly one second later he will do the same, and so on at one-second intervals until his message is complete. Alice will record what she sees during each interval of one second. If she sees a flash, she will record that as a 1, if she doesn't see a flash, she will record that as a 0.

Alice and Bob have another agreement—each 1 or 0 that Bob sends is a response to the same yes/no question, which is—"Is the letter you're sending on the left half of the alphabetic list?"

Bob and Alice agree that yes is encoded as 1 and no as 0. Now, if Alice sees a 1, she discards the right half of the list. If she sees 0, she discards the left. When she sees the second 1 or 0, she halves the list of letters again. She continues this "list halving" until she's left with only one letter, the one that Bob wanted to send.

So, to send the word *help*, Bob starts with a 1.

Alice discards the *right* half of the alphabet, so she's left with *abcdefghijklm*.

Bob sends a 0.

Alice now discards the *left* half of the letter string of *abcdefghijklm*, so she's left with *ghijklm*.

(By agreement, if the list of letters is an odd number, then it's always split so the left half is one letter shorter than the right half.)

Bob sends a 1.

Alice discards the *right* half of ghijklm so she's left with *ghi*.
Bob sends a 0.
Alice discards the *left* half of ghi so she's left with *hi*.
Bob sends a 1.
Alice discards the *right* half of *hi*, so she's left with *h*.

Alice has successfully received the first letter of Bob's message, *h*, which was encoded as 10101. She then returns to the full alphabet list and decodes the second letter of the message.

By this method *e* will be 11010: *l* will be 10001; and *p*, 01100. So *help* becomes encoded as 10101, 11010, 10001, 01100.

Alice and Bob have encoded the word *help* with the answers to twenty yes/no questions transmitted using a light being flashed on or off. This system means that any English letter can be sent as the answer to five yes/no questions. This 1-or-0 answer to each such question Shannon christened a "bit" of information. According to our code, therefore, the informational size of a single letter of English is five bits.

Readers may be aware that the word *bit* has another meaning—it can also represent a single digit of a binary number. So, the number 2 can be represented as the "two-bit number" 10, 3 as 11, 4 by 100, and so on. Shannon's bits are different. They are simply the answer to yes/no questions, and their purpose is to quantify information by counting the number of such questions needed to convey a message. Going back to the way Bob transmitted *help* to Alice, the key lesson is that it needed twenty bits to do so. The specific value of each bit was just an artifact of the conventions agreed to by Alice and Bob.

So, is this conversion of a message into a string of yes/no bits an objective measure of its size? Shannon wanted a way of measuring information that could be universally agreed upon and independent of the encoding method. To do that, he argued that the size of a piece of information must be the *smallest* number of bits needed to encode the message. With the message *help*, is it possible to use fewer than twenty bits to send it? The answer is yes if you account for the fact that some letters appear far more frequently than others in any written English text. In the example above, Alice and Bob treated each letter as equally likely, which is not the case with actual English text. What Shannon showed is that you have to take this statistical pattern into account.

Across all English texts, *e* is the commonest letter, occurring 12.7 percent of the time. *T* is next, occurring 9.1 percent of the time, all the way down to *z*, which as well as being the last letter in the alphabet only occurs 0.074 percent of the time. This is why the game of Scrabble allots far higher scores to the use of a *z* than an *e*. Shannon showed that knowledge of these statistics can reduce the number of bits you need to send a message.

To understand why, imagine that Alice and Bob's alphabet reflects the statistical likelihood of each letter. This "statistically accurate" alphabet is ordered *etaoinshrdlcumwfgypbvkjxqz*. Now amend this alphabet so it has more copies of common letters to reflect the frequency at which they occur in English; i.e., it has 172 *e*'s, 122 *t*'s, 110 *a*'s, 101 *o*'s, 94 *i*'s, 91 *n*'s, all the way down to 2 *x*'s, 1 *q*, and 1 z. In total there will be 1,351 letters, creating an alphabet that looks as follows:

*eeeeeeeeeeeeeeeeeeeeeeeeeeeeeeeeeeeeeeeeeeeeeeeeeeeeeeeeeeeeeeeeeeee*
*eeeeeeeeeeeeeeeeeeeeeeeeeeeeeeeeeeeeeeeeeeeeeeeeeeeeeeeeeeeeeeeeeeee*
*eeeeeeeeeeeeeeeeeeeeeeeeeeeeeetttttttttttttttttttttttttttttttttttttttttttttttttt*
*ttttttttttttttttttttttttttttttttttttttttttttttttttttttttttttttaaaaaaaaaaaaaaaaaaaa*
*aaaaaaaaaaaaaaaaaaaaaaaaaaaaaaaaaaaaaaaaaaaaaaaaaaaaaaaaaaaaaaaa*
*aaaaaaaaaaaaaaaaaaaaaaaaaaaaaaaaaaaooooooooooooooooooooooooooooooo*
*oooooooooooooooooooooooooooooooooooooooooooooooooooooooooooooooooo*
*ooooooooooooiiiiiiiiiiiiiiiiiiiiiiiiiiiiiiiiiiiiiiiiiiiiiiiiiiiiiiiiiiiiiiiiiiiiiiiiii*
*iiinnnnnnnnnnnnnnnnnnnnnnnnnnnnnnnnnnnnnnnnnnnnnnnnnnnnnnnnnnnnnnn*
*nnnnnnnnnnnnnnnnnnnnnnnnnnnnnnnnnnnnnnnnssssssssssssssssssssssssssssssss*
*ssssssssssssssssssssssssssssssssssssssssssssssssssssssshhhhhhhhhhhhhhhhhhhh*
*hhhhhhhhhhhhhhhhhhhhhhhhhhhhhhhhhhhhhhhhhhhhhhhhhhhhhhhhhhhhhhhh*
*hhhhhrrrrrrrrrrrrrrrrrrrrrrrrrrrrrrrrrrrrrrrrrrrrrrrrrrrrrrrrrrrrrrrrrrrrrrrrrr*
*rrrrrrrrrrddddddddddddddddddddddddddddddddddddddddddddddddddddddd*
*dddddddddllllllllllllllllllllllllllllllllllllllllllllllllllcccccccccccccccccccccccccccccccc*
*cccccccuuuuuuuuuuuuuuuuuuuuuuuuuuuuuuuuuuuuuuuuuuuuuuummmmmmmmmm*
*mmmmmmmmmmmmmmmmmmmmmmmmmmmmmmmwwwwwwwwwwwwwww*
*wwwwwwwwwwwwwwwwwfffffffffffffffffffffffffggggggggggggggggggggg*
*gggggggyyyyyyyyyyyyyyyyyyyyyyyyyyyyyyppppppppppppppppppppppppppppppbb*
*bbbbbbbbbbbbbbbbbbbvvvvvvvvvvvvkkkkkkkkkkjjxxqz*

Now, using the same yes/no system as before, Alice will find that the more common the letter, the fewer the number of bits she needs to send it. For example, the letter *e* will only need three bits, 111, to send.

The first 1 will halve the statistically accurate alphabet down to its first 675 letters, which now only includes *e*, *t*, *a*, *o*, *i*, and *n*.

The second 1 narrows the selection down to the first 337 letters, which only includes the letters *e*, *t*, and *a*.

The third 1 narrows the alphabet down to the first 168 letters, all of which are *e*.

So, 111 equals *e*.

We can use this statistical knowledge to shorten the message *help*. *H*, which occurs 6 percent of the time in English, is neither too rare nor too common, so it will still take five bits to communicate. *E*, as we've seen, will only take three bits, 111, to send. *L*, like *h*, will take five bits to send, and *p*, a relatively rare letter, which occurs 1.9 percent of the time, will take six bits to send. So, the total number of bits needed to send *help* is 5 (for *h*) + 3 (for *e*) + 5 (for *l*) + 6 (for *p*), which comes to nineteen bits.

By taking letter-frequency statistics into account, we shortened the message *help* by one bit. It's true that uncommon letters take more bits to send than in the previous system, but because common letters take fewer, overall the number of bits needed comes down. A word such as *heat*, with three common letters (*e*, *a*, and *t*) will only need sixteen bits. Over long messages this method means that on average Alice and Bob's messages will use up around 10 percent fewer bits than if they assumed all letters are equally likely.

As Shannon pointed out, single-letter frequencies are not the only statistical patterns in English text. Many pairs of letters—*th*, *he*, *in*, and *er*, for example—are more common than others. Many others such as *gx* never occur. Sometimes one letter determines the next, as in the case of *q*, which is always followed by *u*. By taking all these patterns into account it's possible to reduce the average number of bits needed to send an English message down to around 1.6 bits per letter. (It may seem odd to refer to a fraction of a bit; 0.6 of a bit doesn't mean 0.6 of yes-or-no or 0.6 of 1-or-0. It means that, for instance, a message 100 letters long would need 160 bits, on average, to send.)

In his paper, Shannon used the example of written English to explain many of his ideas. But the principles this illustrated are widely applicable. The key idea is that the more patterns that can be identified in any piece of information, the less the number of bits needed to encode it.

It's fairly easy, for instance, to see how this applies to images. Intuitively, it makes sense to say that an image of random-colored dots requires many more bits of information—many more yes/no questions—to convey than an image of a repeating pattern such as a set of horizontal stripes. In the former, you would have to specify the color and brightness of every point in the image. In the latter, you'd only have to specify two colors and how often they repeat.

Images in the real world are rarely a random collection of colored dots or a set of regular stripes, but they do contain patterns. Engineers exploit these to reduce the number of bits needed to store and transmit video and still images.

This methodology also applies to the spoken word, which is made up of an alphabet of sounds—some clump together, others are always separate. By identifying these statistical patterns, it's possible to reduce the number of bits needed to communicate comprehensible human speech.

It's in this relationship between statistical patterns and the number of bits needed to convey a piece of information that its relationship with thermodynamics lurks. The reason lies in the form of the mathematical equations Shannon used.

Why? Because the formula he derived for measuring the average number of bits needed to encode a piece of information looked almost exactly like Ludwig Boltzmann and Josiah Willard Gibbs's formula for calculating entropy in thermodynamics.

Here's Shannon's equation for calculating the size of any given piece of information:

$$H = -\Sigma_i \, p_i \log_b p_i$$

And here's one way of stating Boltzmann's equation for calculating the entropy of any given system:

$$S = -k_B \, \Sigma_i \, p_i \ln p_i$$

These two equations don't just look similar; they're effectively the same.

Shortly after deriving his equation, Shannon pointed the similarity out to John von Neumann, then widely considered the world's best mathematician. Von Neumann shrugged, suggesting that Shannon call his measure of the number of bits needed to carry a piece of information *information entropy* on the grounds that no one really understood thermodynamic entropy either.

The similarity occurs because Shannon considered a system of communication like written English in much the same way that Boltzmann had thought about a gas.

Remember again the example of the air in your kitchen. If the heat is concentrated in hot spots—e.g., inside the oven—the molecules here possess more energy, on average, than those in the rest of the room. But there are far fewer ways of achieving such an energy distribution than for the energy to be spread throughout the room. Hence over time, if the oven door is open, the heat will disperse.

Shannon's logic is similar.

The longest nontechnical word in English, *antidisestablishmentarianism*, is a string of twenty-eight letters.

Imagine a large circle, which is proportional in size to all meaningless letter combinations starting with strings of length one, all the way up to strings of length twenty-eight. This is the equivalent of the kitchen where the heat is dispersed.

Next to it is a much smaller circle whose area is proportional to the number of actual English words. This is equivalent to a kitchen with a hot spot.

To send a message accurately using English text, both the sender and the receiver must stay within the small circle. Interference or noise will push the message out into the larger circle of random-letter arrangements. This is akin to the way heat will dissipate out of a hot spot such as an oven, moving from unlikely to likelier arrangements of energy distributions.

One must take measures to preserve a message from degenerating just as one does to prevent heat from dissipating. For the latter, we use insulating materials. For the former, we use an equivalent technique, which Shannon called redundancy. These are letters or words that don't themselves have meaning—they serve to protect meaning from degenerating into noise.

Take the following example, based on one of Shannon's:

MST PPL HV LTL DFCLTY RDNG THS SNTNC

It's twenty letters shorter than the "correct" spelling but contains no less meaning. Shannon estimated that he could resurrect the meaning of about 70 percent of any text after deleting 50 percent of the letters at random.

Spoken language also has many redundancies. Articles such as *the* and *a* can often be dispensed with. Context also makes many words unnecessary. If you hear, "The storm did much . . . ," you're likely to guess the next word is *damage*. Lovers frequently complete each other's sentences because they are aware of the context of what the other says. Speaking to strangers, on the other hand, requires longer sentences to compensate for lack of shared context. Fascinatingly, it seems that human languages evolved to be spoken and written with redundancies because of the many ways segments of a message can be lost—from trying to speak in a noisy marketplace to communicating with children or adults who are new to a language. We put in pauses and repetitions to protect meaning.

Instinctively, we are all aware of this. We add or subtract redundancy to our messages in response to the amount of interference they need to overcome just as we add or remove layers of clothing in response to the outside temperature. When we send a text message, we are confident that the letters will transmit without loss and that our reader understands the context. This is a noise-free communication channel, so we remove many redundant letters—e.g., we type, "c u ltr at pb."

By way of contrast, we add redundancy when speaking over a distorted phone line—"MY . . . NAME . . . IS . . . PAUL. That's $P$ for Papa, $A$ for Apple, $U$ for Umbrella, $L$ for Lima."

Linguistic redundancy prevents the ideas you wish to communicate from degenerating into noise. Just as some heat must be lost while it flows from hot to cold, creating work, some words and letters must be lost or garbled as a message travels.

This understanding of information entropy and redundancy is why we can build the data networks. Take services such as YouTube or Netflix that hold and distribute huge files of video information. These companies reduce the number of bits that make up these files to be as close to their Shannon entropy as possible. This is called compression, and if it weren't done, the files' sizes would be too large for our networks. The companies that maintain these networks then add digital redundancy to the compressed files to protect them from noise. These extra bits are a sophisticated electronic version of spelling out a word for clarity over a distorted phone call.

Traveling over distance isn't the only way information is lost. Meaningful information tends to degenerate over time, too. Humans have long understood this. Ink fades, paper yellows, and tears and the inscriptions on clay or stone are prone to weathering. We've combated this with permanent inks and durable parchments, but even these are lost when libraries burn down. So we add redundancies by making multiple copies of the texts we consider important, often in many languages. The writers of the Rosetta stone added redundancy by inscribing the same message in three languages, thus demonstrating how this strategy can overcome two thousand years of decay. Written languages are themselves a form of redundancy to protect against the loss of information over time. They add no meaning to the spoken word; they exist to protect the meaning of those words long after the brains that first conceived them have turned to dust.

The scientist Rolf Landauer captured this idea with the phrase "information is physical." All forms of information require a change in the physical universe. Written words require marks to be made on a physical medium of some kind. But even a spoken word requires the movement of vocal cords, which make air molecules vibrate. Similarly, a thought requires electrochemical changes in the neurons in our brains. In this sense, information entropy is constrained by thermodynamic entropy. As physical systems decay, so does any information they carry. Imagine writing your name in the sand on a beach. This act arranges the sand particles into an unlikely, low-entropy configuration, a pattern that has meaning. When a wave arrives, that meaning is lost as the sand particles become jumbled and are rearranged into more likely but less meaningful configurations—a high-entropy state.

However we choose to record information, the relentless increase of entropy will erase it as surely as a wave wipes away a name drawn in the sand. William Thomson's prediction of the heat death of the universe includes thoughts, words, and memories. Everything will end up at the same temperature, everything will be forgotten.

In July 1948, when Shannon published his paper, no one predicted the scales at which his ideas would be applied. Nor did anyone question whether the similarities between thermodynamic and information entropy are coincidental, or if they are two faces of the same phenomenon.

What changed that was another Bell Labs discovery. For on June 30, 1948, just a few days before Shannon's paper appeared, a department within the same organization specializing in "solid-state physics" held a press conference in New York City. There they unveiled a strange device, the size of a cob of corn, which had three wires sticking out of it. Alongside it was a human-size replica to give the invited journalists a clearer sense of what the tiny object being demonstrated looked like.

The Bell Labs' latest invention was the transistor. It's a grand irony that none of the three uses of the transistor demonstrated at this most historic of all tech demos showed off its most useful feature. Instead the Bell Labs engineers pitched it as a smaller and more reliable replacement for the thermionic valve—that is, as a device that could amplify analog signals.

To demonstrate this capability, each audience member was given a pair of headphones with which they listened to transistor-amplified voices, a transistor-enabled radio broadcast, and a transistor-generated whistling sound. Little reference was made to how the transistor can also act as a tiny, low-power on-off switch. In other words, it's perfectly suited to answer yes/no questions and unleash the power of the bit.

Once this fact was appreciated, engineers focused their efforts on miniaturizing the transistor, and as a result, they have shrunk in size and grown in number at exponential rates. Today's microchips can cram up to 20 billion transistors into the space that first, single transistor occupied, which means each individual transistor is around a millionth of a millimeter across. One estimate suggests that between the time the technology was invented and 2014, around 3 sextillion transistors have been built— that's 3 followed by twenty-one zeros. By way of comparison, a mere 2 hundred billion—2 followed by eleven zeros—stars are in the Milky Way. Each of these transistors answers billions, if not, trillions of yes/no questions every second so we can inform, insult, entertain, and communicate with one another or do whatever we need to do with information.

In the modern world, the thermodynamic cost of information is determined by the electrical properties of silicon. Typically, every time a single transistor switches between on and off to answer a yes/no question, it dissipates around 10 million-millionths of a joule of heat into its surroundings. That's a small amount. But consider that in a single chip 10 million transistors are switching on and off a billion times a second. That means that a chip with a surface area of one square centimeter can easily dissipate heat at a rate of several tens of joules per second (i.e., tens of watts). If left uncooled, the chip's surfaces would become hotter than the hot plates on stoves. Even more extraordinary is what will happen if transistors continue to shrink for the next two or three decades at the same rate at which they have since the 1960s. Every square centimeter of the surface of a chip built of such transistors will give off heat at a rate of one thousand kilowatts. That's similar to the rate at which heat is given off in the exhaust nozzle of a rocket, and only a few times less than the six-thousand-kilowatt rate at which heat is given off on the surface of the sun.

Clearly that isn't going to happen; we simply cannot build cooling systems that can safely remove that amount of heat in our computer systems. Any chip built with such a high density of transistors would melt as soon as it was turned on. And because of the laws of heat, that means we are not far from reaching the end of the road that conventional silicon transistors have taken us down. Unsurprisingly, there's been burgeoning growth in research and development to see if we can reduce the amount of heat dissipated every time a bit is processed.

That we humans generate large amounts of waste heat when processing information should not be seen as a retrograde step. Quite the opposite. After all, processing bits allows us to perform tasks much more efficiently than in the pre-digital age. Take electronic books. Distributing this book as bits over the internet to you uses far less energy, dissipates far less heat, and creates far less carbon than if this book is printed on paper and shipped in trucks, ships, and planes. A study by the think tank the Climate Group estimates that, although by 2020 the digital sectors of the world's economy may be creating in the region of 1.43 billion tons of carbon every year, they may also be reducing the carbon created by the rest of the world economy by nearly 8 billion tons per year, a net saving of over 6 billion tons of carbon per year.

The pressure is on us, then, to move more of our activity from the physical to the digital sphere, while striving to make our digital machines as efficient as possible. But will we find, as we did with heat engines, that thermodynamics limits the extent of our improvements? Must we always dissipate heat when we process information? Would it be possible to build a machine that can process information—and even think—without this cost? Without hastening the end of time?

For an answer, scientists have had to revisit a thought experiment first elucidated in the 1860s by a scientist we met several chapters ago: James Clerk Maxwell.

And in so doing, they had to resurrect a demon.

# CHAPTER SEVENTEEN

# Demons

The process of diffusion could be perfectly prevented by an army of Maxwell's intelligent "demons."

—William Thomson

We last met James Clerk Maxwell in the late 1860s in his London garret, where, with his wife, Katherine, he performed experiments to test his statistical description of the behavior of gas molecules. In the years that followed, Maxwell's scientific interests turned to electricity and magnetism, resulting in his magnum opus, a set of equations describing the behavior of electricity and magnetism with great precision. They led to the discovery of radio and paved the way for Einstein's theory of relativity. Maxwell retained a keen interest in thermodynamics, however, and his reputation was such that when a friend, the physicist Peter Guthrie Tait, planned to write a history of the field in 1867, he wrote to Maxwell for insight and assistance.

Tait was British, and his motives were partly nationalistic. He had already published a brief history of thermodynamics under the title "Historical Sketch of the Dynamical Theory of Heat" in the journal *North British Review*. However, that had attracted the ire of Rudolf Clausius, with some justification. Tait had given the greater part of the credit for breakthroughs in thermodynamics to British scientists such as William Thomson and James Joule and barely mentioned their European counterparts. Tait had therefore written to Maxwell, saying, "Clausius and others have cut up very rough in bits referring to them" and hoping for his friend's support.

In his reply, Maxwell politely declined to become involved in the row, commenting, "I could make no assertions about the priority of authors." He did, however, agree to contribute sections that would clarify the science and point out any potential flaws in current thermodynamic theory. To that end, Maxwell came up with a thought experiment that has become legendary in the history of science. This was also the first time anyone had conceived of a possible link between energy, entropy, and information, and it has engendered fruitful scientific discussion for well over a century. The thought experiment is now referred to as Maxwell's demon.

In his letter to Tait, Maxwell's aim is "to pick a hole in the 2nd Law of Thermodynamics; that if two things are in contact, the hotter cannot take heat from the colder without external agency." This law, discovered thanks to the combined efforts of William Thomson, Rudolf Clausius, and others, had by the 1860s become regarded as universally true. It also chimed with people's intuition and experience—heat never flows spontaneously from a cold object to a warmer one. After all, a cup of lukewarm tea never becomes hotter of its own accord by drawing heat from the cold table it is sitting on.

Maxwell sought to challenge the inevitability of this observation with a thought experiment. He wanted to show that under certain, admittedly unusual, circumstances, heat might flow the "wrong" way, from cold to hot without a compensatory flow in the other direction somewhere else. And strangely, it seemed that the way to achieve that was with the use of information.

Maxwell asks us to imagine a sealed box that's full of gas. The box is divided within into two equal parts by a diaphragm—a thin partitioning wall that does not permit any gas molecules to flow across it. Maxwell then stipulates that the temperature of the gas on one side of the diaphragm is hotter than on the other side, and he reminds us what this means at a molecular level: the average speed at which the molecules are moving on the hot side is greater than the average speed at which they move on the cold side. But, as Maxwell had shown, these are average values. So, some molecules on the hot side of the diaphragm will be sluggish and move at a lower speed than the average speed of the molecules on the cold side. Similarly, some molecules on the slow side will be moving at higher speeds than the average speed of the molecules in the hot half of the box.

Now comes an insight as startling as it is playful. "Conceive a finite being," writes Maxwell, "who knows the paths and velocities of all the molecules by simple inspection but who can do no work except open and close a hole in the diaphragm by means of a slide without mass."

Maxwell's "finite being" can open or close a sliding door in the diaphragm that divides the box. The "slide without mass" is important because it implies that moving it requires no energy. That in turn means the "being" needs no energy to open or close the door. What it does need, crucially, is the ability to obtain *information* about the individual molecules on both sides of the partition. It observes the speeds at which they move. In particular, it pays close attention to molecules that, by chance, approach the doorway in the partition that divides the box. The finite being is most interested in an unusually sluggish molecule on the hot side, one that is moving at a *slower* speed than the average speed of the molecules on the cold side. When the being sees one of these approaching the door, it slides it open. As a result, the sluggish molecule drifts out of the hot side of the box into the cold side.

Similarly, the finite being looks out for a speedy molecule on the cold side—for any molecule there that is moving at a *higher* speed than the average speed of the molecules on the hot side. When it sees one approaching the doorway, it opens it, and the speedy molecule moves from the cold to the hot side of the box.

After a while, Maxwell concludes, something extraordinary will have happened. More and more speedy molecules will accumulate on the hot side of the box, and more and more sluggish molecules will accumulate on the cold side. This will mean that the cold side will have become colder and the hot side hotter. This stands in direct opposition to the second law of thermodynamics, which explicitly states that heat cannot flow from hot to cold without the expenditure of work. Yet here, Maxwell points out, "No work has been done, only the intelligence of a very observant and neat-fingered being has been employed."

Maxwell made no attempt to describe how his finite being could operate without energy, nor how a sliding door of "no mass" could be built. Such ideas were fanciful and unrealistic, and Maxwell's intention with this thought experiment was to show the validity of the second law. He argues later in the same letter to Tait that, yes, if we could detect and exploit the movements of individual molecules, we could reverse the second law. But in reality, such observations would be impossible to achieve. Or as he puts it, "We can't, not being clever enough."

In 1871, four years after posing the thought experiment in his letter to Tait, Maxwell described it again in a textbook entitled *Theory of Heat*. Shortly after, the idea appears to have caught the eye of William Thomson, who in an 1874 paper published his own thoughts, describing Maxwell's "finite being" as a "demon." The moniker stuck. Thomson, like Maxwell, stressed the absurdity of the demon to make the point that in the real world, one without demons, the spontaneous flow of heat is always from hot to cold and never the other way. The second law of thermodynamics is safe.

For the next six decades, Maxwell's demon lived in relative obscurity. Then in 1929, it returned to tantalize us with a possible link between information, energy, and entropy. This time, it was resurrected by Leo Szilard, the scientist whom we met in chapter 15.

In 1929, Szilard was living in Berlin and working with Einstein on their designs for safer refrigerators. He had also, in his PhD thesis, investigated the statistical underpinnings of thermodynamics. So Szilard had a deep understanding of the subject from both a theoretical and practical perspective. Now Maxwell's demon captured his imagination. But where Maxwell and Thomson had seen the demon as a way of testing the second law of thermodynamics' validity, Szilard thought it could provide insight into the physics of information.

Szilard simplified the task that Maxwell had set his demon. In the original thought experiment, the Scotsman had surmised that the demon would have to make countless measurements of the speeds of many different molecules to bring about its reversal of the second law of thermodynamics. However, in his own paper, brilliantly entitled "The Decrease of Entropy by Intelligent Beings," Szilard argued that such an immense task was unnecessary for the demon to work his mischief.

Like Maxwell, Szilard asks us to imagine a box with a partition. Inside this box, however, there is only one molecule moving about. Initially, it's free to move throughout the box, now and again colliding with and bouncing off the walls. Szilard's demon has therefore to watch only one molecule, whereas Maxwell's demon had trillions. Szilard then simplifies his demon's task even further. He asks only that it observes which half of the box the molecule is in—left half or right half—at any given time. Once it, say, sees that the molecule is in the left half of the box, Szilard's demon brings down a partition that divides the box in two, effectively trapping the molecule in the left half of the box.

Szilard stipulates that the partition be movable—it can slide back and forth inside the box the way a piston can slide back and forth inside the cylinder of an engine.

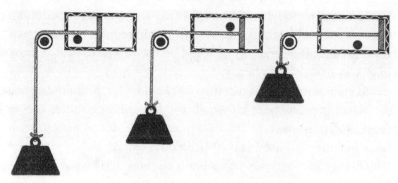

An engine driven by a single molecule

Once the demon knows which side of the box the molecule is in, it acts. If it knows the molecule is in the left half of the box, it attaches a weight using a pulley to the left side of the sliding partition. Now as the particle bounces around, it occasionally collides with the movable partition. This action pushes it toward the right, and as the partition moves, it lifts the weight.

The point of all these contrivances is that, with one simple piece of information, namely the knowledge that the molecule is on the left side of the box, Szilard's demon has been able to lift a weight—i.e., do work. The demon can repeat this process ad infinitum. It is effectively getting something for nothing simply thanks to a single bit of information—left or right. I've used the word *bit* here deliberately. Left or right is a binary choice just as 1 or 0 is. With binary information alone, Szilard's demon appears to be able to have converted the random movement of a molecule into useful work. This contradicts the second law of thermodynamics by implying that we wouldn't need heat to flow from hot to cold to do useful work. With "Szilard's demon," we could obtain power from any volume of gas, even if it was at a constant temperature throughout. Indeed, if enough of these "Szilard demons" were unleashed, we could generate all the electricity we needed from the air in the earth's atmosphere! It seems possible to construct "a perpetual-motion machine," as Szilard puts it, simply if "one permits an intelligent being to intervene in a thermodynamic system."

So what does this mean? We saw in the previous chapter how processing information increases entropy. Is Szilard, with his thought experiment, suggesting that information can do the opposite, overcoming the second law of thermodynamics and turning warm air at a constant temperature into useful work? Such a system would reduce the entropy of the universe because this "free" work could be used to force heat to flow the "wrong" way from cold to hot.

Szilard emphatically states that this can't happen for the following reason:

The act of measurement by which the demon determines the molecule's position must cause an increase in entropy that compensates for any decrease in entropy caused as the piston does work.

Something about Szilard's argument is circular, and he was vague about how his demon causes entropy to increase. But his paper is the first to claim that processing bits of information must dissipate heat or else we could build perpetual motion machines in defiance of the laws of thermodynamics. What's remarkable, too, is that Szilard wrote his paper in 1929, decades before bits were used in global communication networks and before people understood how important they were to the transmission and storage of information.

Over the next three or so decades, the Szilard/Maxwell demon lurked in the shadows; scientists regarded it as an intriguing puzzle but one of little relevance. The few scientific papers over this period that did mention the demon followed Szilard's reasoning. They speculated on what kind of apparatus a demon might use to measure the position of a molecule and concluded, much as Szilard had done, that any such system will dissipate enough heat to balance the entropy decrease resulting from the piston as it lifts a weight.

From the 1950s onward, however, as the number of bits and transistors in the world began to proliferate and computers began to generate appreciable quantities of heat, the Maxwell/Szilard demon went from academic curiosity to being technologically and commercially relevant. Scientists at the computer giant IBM's research department revisited the demon as the question of whether information has a thermodynamic cost acquired a new urgency. And just as over a century earlier Sadi Carnot had realized that to properly understand a steam engine one must look beyond its engineering practicalities to its underlying principles, IBM's scientists realized that to properly understand information, the machines that manipulate it must also be idealized and considered in the same way.

As two IBM researchers in the field, Rolf Landauer and Charles Bennett, wrote of their work, "We are looking for general laws that must govern all information processing, no matter how it is accomplished. Any limits we find must be based solely on fundamental physical principles, not on whatever technology we may currently be using."

The senior figure of the two, Rolf Landauer, was born to a Jewish family in Stuttgart in Germany on February 4, 1927. His father, Karl, a successful architect and builder, died in 1934 from wounds incurred fighting for Germany in World War I. Convinced to the end that the Nazis were a short-lived fad, Karl's last letter to his wife, Anna, asked that she raise their sons as good Germans. Anna, however, grasped the nature of the Third Reich and emigrated with her family to New York early in 1938. Rolf excelled academically in his adopted country, graduating from Harvard in 1945. He then joined the US navy, training as an electronic technician's mate. Landauer credited this practical experience as invaluable to his later work.

Despite his stellar academic credentials, Landauer found many American universities and industrial laboratories were reluctant to take on Jewish employees in the early 1950s. Encouraged by an old friend, he joined IBM's Research Laboratory in 1952, then newly established on the site of a disused pickle factory in Poughkeepsie in New York State. As AT&T management had, IBM boss Thomas J. Watson Sr. encouraged his researchers to pursue their scientific interests without commercial pressure. Additionally, the IBM lab fostered close ties with scientists at universities such as Columbia.

Landauer arrived at IBM during a significant time in the history of the computer—he witnessed how they changed from using thermionic valves to transistors. Computers are, in essence, giant arrays of on/off switches. Thermionic valves performed this role in early machines, but they were power hungry, unreliable, and large—roughly the size of a light bulb. An early machine, ENIAC, funded by the US army to calculate artillery firing tables, occupied eighteen hundred square feet, weighed twenty-seven tons, and used 174 kW of electricity. ENIAC chucked out heat— two 20 horsepower fans blew cool air at it to prevent it from overheating.

The transistor, which Bell Labs invented in 1948, worked as a switch as well, but it was the size of pea. It also consumed a fraction of the power and gave off far less heat than the thermionic valve. That meant that when IBM brought out its first transistor-based computer in 1958, it offered striking advantages over its valve-based predecessor. Faster and more powerful, it weighed half as much. Power consumption for the computer and its air-conditioning unit were down by over 60 percent. To engineers and scientists working in the field, this proved that miniaturization was the future—the smaller the transistor, the more that could be crammed into a given space, and that in turn meant more computing power.

With remarkable prescience, Landauer began to investigate where miniaturizing electronic components might lead. As he put it in a 1961 paper, "The search for faster and more compact computing circuits leads directly to the question: What are the ultimate physical limitations on the progress in this direction?"

Then in 1972, twenty-nine-year-old Charles Bennett joined Landauer at IBM Research. Originally trained as a chemist, he then received a PhD from Harvard for work that used computer simulations to illuminate molecular behavior. Together, he and Landauer deduced the ultimate thermodynamic cost of a bit.

To see how, picture again Leo Szilard's demon who uses the information about the position of a single molecule inside a box to generate work. Now imagine the demon can measure the particle's position with an apparatus that is so perfectly manufactured that it works without dissipating any heat. This might seem to render the argument pointless, but it's no different from Sadi Carnot's approach when he asked his readers to consider a steam engine that operated without friction.

Let's start by considering what happens when the particle is in the left half of the box. The demon observes this as a "bit" of information, which it acts on by attaching a weight to the piston. The particle strikes the piston, as I described above, which moves and lifts a weight.

But what happens after the partition has been pushed fully to one side of the cylinder? How does the demon ensure that the extraction of work from the molecule continues?

It must repeat the process above—to do that it reinserts the partition in the middle of the cylinder and obtains a second bit of information by observing where the molecule is now. As before, it attaches a weight to the partition and lets the molecule push it.

But a problem lurks here—namely the previous bit of information. The demon must erase that to make room for the new bit. But, wait, perhaps the demon has a large memory-storage device? Even so, at some point that will fill up, and for the demon to carry on, it will have to start erasing the earlier bits of information it acquired.

Here lay the answer to the mystery of the minimum thermodynamic cost of a bit. Landauer and Bennett's point was that for the demon to continue its work, it must at some point start erasing bits of information. It must forget earlier measurements to make room for new ones. And this act of forgetting must dissipate heat by an amount that compensates for the work done by the moving piston.

Think again of Sadi Carnot's description of a steam engine. He argued that useful motive power such as lifting a weight can only be achieved by a steam engine if heat flows from a hot source such as a furnace to a cool "sink" such as the atmosphere. The key feature of this sink is that it must be able to absorb any amount of heat without becoming noticeably warmer. That's a realistic assumption because real engines dump heat into the earth's atmosphere, which doesn't immediately become warmer. Now imagine what would happen, however, if the cool sink did not have a limitless capacity to absorb heat. It would slowly become hotter as the heat from the furnace flowed into it. After a while, the sink would reach the same temperature as the furnace, and at that point the engine would stop working. It would do no more work even if fuel continued to burn in the furnace.

Landauer and Bennett showed that the flow of information is analogous to the flow of heat. Just as for a steam engine to work it must discard or dissipate heat, so, too, the demon must discard bits. As it discards each bit, some heat will dissipate from its memory, irrespective of what material or mechanism has been used to store that bit.

You could argue that if the demon's memory were infinitely large, it could store all the "used" bits and do work without ever dissipating heat. That's true in principle but not in practice. In reality, just as a steam engine whose sink becomes as hot as its furnace will stop working, so, too, a demon whose memory becomes full of "old" bits of information will stop working. For it to start up again, it will have to start erasing those stored bits so new information can "flow" into its memory.

The extraordinary thing about this line of reasoning was that it allowed Landauer and Bennett to calculate the amount of heat dissipated when erasing a bit of information, *even if the means of acquiring and storing the bit is frictionless.* I mentioned earlier that a real transistor dissipates about a 10-million-millionth of a joule every time it switches from off to on or vice versa. Most of that heat arises as subatomic particles move within the silicon of which the transistor is made. But let's assume that the demon's memory is made of perfect transistors that dissipate no heat at all. Even so, when it discards a bit of information, a small quantity of heat will be dissipated. This quantity is therefore the minimum amount of heat dissipated when a bit of information is erased.

This amount is a fundamental limit set by the laws of physics, as fundamental as the law that tells us nothing can travel faster than light. Today it's called the Landauer limit, and it tells us that however good the technology we create to process bits, once we start erasing those bits, we will make the environment a little warmer. By how much? At temperatures common on the earth's surface, the amount of heat that will be dissipated when even a perfect storage device erases a bit of information is 3,000 billion-billionths of a joule.

This limit has been confirmed in physics laboratories around the world in the years since 2012. Among the first scientists to do so were Eric Lutz and his colleagues at the University of Augsburg in Germany. That means we have an answer to the question posed in the last chapter: In principle could we build a machine that could think without increasing the entropy of the universe? No, albeit with one caveat.

There is an intriguing possibility that if a computer could be built that didn't need to erase data, it wouldn't dissipate energy. Such a machine wouldn't necessarily need an infinite memory, but in some sense, it would be able to recall all the historical data it once held. It's rather like building a car that loses no energy to friction and brakes by charging a battery. To speed up again, it reuses that stored energy. If these back-and-forth energy transfers are perfect, in theory, the car could run forever without refueling. In that vein, one can envisage a computer that can reverse all the steps it has taken and so never forget its past. But just as with the car example, building such a device presents enormous technical challenges. For the foreseeable future, the Landauer limit remains in place.

The Landauer limit, though, is a tiny amount. Real transistors dissipate 10 billion times as much heat. But knowing this ideal minimum heat dissipated when a bit is erased is invaluable because it tells us that the laws of physics permit us to do considerably better than our current silicon-based technology. We may never build a useful computer that erases bits at a rate of heat dissipation as low as the Landauer limit, but knowing it tells us that we can, in principle, reduce the heat coming off our chips by factors of thousands if not millions.

The other reason to believe it is possible to process bits at a far smaller thermodynamic cost than our current technology comes from applying

Claude Shannon's methods of measuring information to a system that has been processing information as efficiently as possible for billions of years—life.

Take, for instance, the humble *Escherichia coli*, a tiny rod-shaped single-cell species of bacteria. We know them as *E. coli* for short. Each is about two-thousandths of a millimeter long and a tenth of that wide. Millions live in our lower intestines, as they do in similar organs of most warm-blooded organisms. In recent years, by studying the chemical processes in *E. coli*, scientists have estimated how many bits of information a single *E. coli* cell must process to reproduce itself. By measuring how rapidly the cells reproduce and how much energy they consume, they estimate an *E. coli* uses ten thousand times less energy to process a bit of information than the transistors used in most human-built information-processing devices.

It's a humbling thought that an organism that lives in our guts can process information far more efficiently than our most intricate silicon transistors. It's extraordinary, though, that by combining our understanding of heat and information we obtained a new insight into the living world. It's as though life exists in the overlap between thermodynamics and information. To understand this new domain, we must return to a character whom we last glimpsed drinking tea with Claude Shannon in the Bell Labs cafeteria, a man whom Shannon himself described as possessing "a great mind, a very great mind."

# CHAPTER EIGHTEEN

# The Mathematics of Life

*. . . a mathematical model of the growing embryo will be described . . .*
*—Alan Turing*

Since the mid-nineteenth century, scientists such as Hermann von Helmholtz had become convinced that living beings, like everything else in the universe, obey the laws of thermodynamics. By the mid-twentieth century the details had been figured out. Scientists understood that a plant takes free energy from sunlight and uses that to capture or "fix" carbon from the atmosphere. They understood, too, how animals take free energy from foods such as sugar to fuel their metabolisms.

Also, by the 1950s, the idea of genes was firmly established. In the cells of all organisms, we now knew, are inherited instructions, a blueprint that guides the formation of each organism.

What remained mysterious was how those genes worked, particularly in the developing embryo. After all, when cells first form, they are all identical, and all contain a complete set of the organism's genes. But how, as they multiply, do those identical cells know which should then turn into stomach cells, brain cells, limbs, and so on?

That Alan Turing played a role in solving this mystery may surprise. He is best known for his pivotal role in deciphering the encryption system used by the German military in World War II. In its early years, Britain's greatest threat came from Germany's U-boats, which devastated Atlantic shipping, cutting vital supply lines from America. The German navy used the strongest form of encryption they had to communicate with their submarines. Turing made crucial contributions to cracking this cipher in the early months of 1941. By June of that year, the British were using this information so successfully that for twenty-three consecutive days that month, U-boats in the Atlantic did not spot a single convoy. "There should be no question in anyone's mind that Turing's work was the biggest factor in the Hut 8's success," wrote Hugh Alexander, one of Turing's fellow code breakers. "In the early days, he was the only cryptographer who thought the problem worth tackling."

This part of Turing's story has been rightly celebrated in biographies, plays, documentaries, and feature films. Far less known is his seminal contribution to developmental biology, which he made by uncovering a beautiful aspect of the second law of thermodynamics.

Alan Mathison Turing was born in London on June 23, 1912, the son of an Englishman who worked as a magistrate in the city of Madras (now Chennai) in India, which at the time was part of the British Raj. Sara Turing, Alan's mother, had traveled back to England to give birth at a time of growing political unrest in India. After Alan's birth, his mother returned to Madras to join her husband, leaving Alan and his older brother, John, to be brought up by a foster family in the town of Hastings on the south coast of England. During the first eight years of his life, Alan Turing saw his parents two or three times when they came home on leave. Fostering one's children in this way was common among the colonial classes—"accepted procedure for those who served the British Empire," as Turing's brother described it.

Alan Turing never referred to this experience. Yet despite his lonely childhood, signs of his eccentricity and genius began to show. His primary school headmistress said of Turing when he was aged nine, "I have had clever boys and hardworking boys, but Alan is a genius." When he moved on to secondary school, a boarding school in the town of Sherborne in southern England, he read, understood, and wrote a description of Albert Einstein's theory of relativity for his mother, so that she, like him, might appreciate the wonders it contained.

Along with mathematics, Turing developed a love for the natural world. At eight, for instance, he began writing a book called *About a Microscope*. In Scotland, during one of his few family holidays, he spent his time observing the flight of bees. At ten, he then became fascinated by a book called *Natural Wonders Every Child Should Know*, which examines the growth of biological systems. Within were descriptions of creatures such as starfish and sea urchins, along with an honest declaration of how poorly understood the systems that underpin life were in those days. All living creatures are made of cells the author stated, but, "How they find out when and where to grow fast, and when and where to grow slowly, and when and where not to grow at all, is precisely what nobody has yet made the smallest beginning at finding out." Thirty years later, Alan Turing would start us on that journey. Meanwhile his mother noticed his early fascination with nature. In a picture she drew of her son during a hockey match at his primary school, the rest of the children are depicted busily engaged in the game, while Alan is far removed from the action, bent over at the edge of the field, staring intently at a clump of daisies. "Hockey, or Watching the Daisies Grow," Sara Turing titled it.

In 1931, aged nineteen, Turing went to King's College Cambridge to study mathematics. He graduated three years later with a first-class degree and was chosen to be a "fellow" of the college, which meant an annual stipend of £300 (worth around £11,000 [$14,000] today) and the freedom to pursue his many mathematical interests. During this time Turing wrote the paper that, along with his wartime work as a code breaker, is the achievement he's best known for. Published in 1936, the paper has the daunting title "On Computable Numbers, with an Application to the Entscheidungsproblem." The Entscheidungsproblem is a mathematical challenge that had been stated in its modern form in 1928 by David Hilbert, Emmy Noether's mentor at the University of Göttingen. In simple terms, the problem asks if there is an automatic way of determining if any mathematical statement is true. For instance, take a statement such as "prime numbers appear at random." A quick test that said, "No, that's false," would save the futile effort of trying to prove it. Similarly, if the test said it's true, it would make that effort worthwhile. Turing's singular and brilliant answer to the Entscheidungsproblem was to conceive of something called a Universal Machine. This machine can be programmed to solve any mathematical problem that a human could solve. Put another way, he came up with the idea that a single piece of hardware could be repurposed to perform a vast multitude of tasks simply by altering the software that it runs. Turing then showed that such a machine would not be able to test the validity of all possible mathematical statements, thus answering Hilbert's question no. Most historians now regard Turing's Universal Machine as the foundational principle underpinning the modern computer.

After a two-year stint at Princeton University, Turing returned to Cambridge to join its mathematics department. But it was now 1938, and even this cloistered community had been affected by the rise of Nazism in Germany. Turing was not impervious, for in the buildup to war, he heard from a Cambridge friend, a brilliant linguist and classics scholar named Fred Clayton, of the plight of Jewish children whom sympathetic groups had helped to flee Nazi-controlled Europe. Many were staying in a Butlin's holiday resort repurposed as a temporary refuge in Harwich on the east coast of England.

Turing wanted to do what he could to help these children. So, on a wet Sunday in February 1939, he and Clayton cycled the fifty or so miles from Cambridge to Harwich. Here Turing met a fifteen-year-old Jewish refugee named Robert Augenfeld. His parents had secured him a place among several hundred children on a train out of Vienna that had been funded by a British Quaker organization. Augenfeld had been at the camp for several months by this time, and Turing offered to become his guardian and to sponsor his education. Augenfeld gladly accepted, little knowing that the Englishman would have struggled to pay any school fees out of his modest Cambridge stipend. In the end, a boarding school in Lancashire named Rossall agreed to take some of the refugee children without a fee. Turing, nonetheless, remained responsible for Augenfeld and would in the following years play an active role in helping him through school and university. Turing took him on holidays and invited him to Cambridge on several occasions. Augenfeld, who never saw his own parents again, never forgot Turing's kindness, and the two men remained friends till Turing's death.

Within a few weeks of meeting Augenfeld, Turing began his code-breaking work at Bletchley Park. He never stated if meeting the refugee children affected his work, but there is no doubt that as the war began Turing knew, in a personal way, what kind of regime the Nazis had created. Turing is often portrayed as awkward and incapable of empathizing with other people, and according to his brother, Alan hated what he called "vapid conversation." But his actions speak strongly of a deep sympathy for his fellow humans. Over the first eighteen months of World War II, Turing did his historic work as a code breaker. During this time he also visited Bell Labs in New York City, where he assessed the cryptographic potential of the SIGSALY voice-encryption system and met regularly for tea with Claude Shannon, one of the few people in the world who could be considered his intellectual equal.

Although the war dominated Turing's attention at Bletchley Park, people noticed his fascination with the natural world and the patterns and shapes it manifests. This interest was shared by Joan Clarke, a fellow mathematician to whom Turing became close and was briefly engaged. Clarke, who had a double first from Cambridge in mathematics, was the most senior female cryptographer at Bletchley Park and worked with Turing on cracking the German naval codes. Marriage to Clarke, Turing felt, would bring with it a certain respectability. Honest to a fault, however, he told her of his "homosexual tendencies." Nonetheless, Clarke agreed, and the couple introduced each other to their respective families. A few months later, however, Turing broke off the engagement, unable to countenance a sham marriage. The pair remained close friends, though, and as a mathematician who had also studied botany, Joan Clarke often accompanied Turing on walks around the grounds at Bletchley Park, helping him identify plant species that particularly intrigued him.

Turing and Clarke were often spotted during breaks from code breaking lying on the grass, peering, in an echo of Sara Turing's drawing, at the spirals of florets at the center of daisy flowers.

The head of the daisy, the circular part to which the flower's petals are attached, consists of densely packed dots called florets, which eventually become the plant's seeds. Close examination reveals that the florets are arranged in spirals that curl in clockwise and counterclockwise directions outward from the center of the flower. What fascinated Turing and Clarke was that the number of clockwise and counterclockwise spirals always form a pair of numbers from a so-called Fibonacci sequence. Named after a twelfth-century Italian mathematician, the series is defined by each subsequent term's being the sum of the previous two terms ($1 + 1 = 2$; $2 + 1 = 3$; $3 + 2 = 5$; $5 + 3 = 8$; $8 + 5 = 13$; $13 + 8 = 21$; $21 + 13 = 34$; and so on). So, in a daisy flower, there might typically be twenty-one clockwise spirals and thirty-four counterclockwise spirals, or fifty-five clockwise spirals and eighty-nine counterclockwise ones. These Fibonacci sequences can be found all over the material world. Take fir cones, for instance, the seedpods of which are usually arranged as a set of clockwise and counterclockwise spirals, and just as with the daisy, the numbers of each type of spiral are a pair of Fibonacci numbers. A top-class long-distance runner—his marathon time was two hours and forty-six minutes—Turing would often return from runs with fir cones to show his fellow code breakers.

Toward the end of the war, as demand for Turing's cryptographic skills decreased, he began to consider real machines that might behave like the theoretical Universal Machine he'd conceived back in 1937. These would be devices that could be programmed to perform many different mathematical tasks—in other words, a computer. When the war ended, a British government laboratory, the National Physical Laboratory, the NPL, based in Surrey, agreed to help realize this vision, and Turing went to work there in October of 1945. However, he fell out with the NPL's engineers, who considered his plans too ambitious. Disgruntled, he returned to Cambridge for sabbatical. Nonetheless, the NPL did build the Pilot ACE, a scaled-down version of Turing's design.

With time to think, Turing focused on what he felt was the fascinating confluence of mathematics, computing, and biology. From 1947 to 1948, he wrote groundbreaking papers on the way nerve cells in the brain might work and on how machines might emulate that process. Then in 1948, Max Newman, also a code breaker from Bletchley Park and now professor of mathematics at Manchester University, recruited Turing. Newman had also secured funding to research and build computers and knew, rightly, that Turing's expertise would be invaluable. The machines that Turing and his colleagues then developed in Manchester were vast in size and limited in power. The first version, known as Baby, weighed one ton and could carry out little more than basic arithmetic. Turing himself wrote the code that enabled Baby to do long division. But Baby and its successors pioneered concepts such as random-access memory (RAM) that are essential to all modern computers. Turing played a crucial role by testing these machines' capabilities by running ever-more-complex software on them.

Inspired by this firsthand experience of the world's first computers, in 1950 Turing wrote a now-famous paper in the philosophy journal *Mind* entitled "Computing Machinery and Intelligence." In it he presented a series of arguments in favor of the idea that machines would one day be able to think as well as or even better than humans. In this paper he introduced "the imitation game," the idea that if a computer provides answers that are indistinguishable from those that a human might provide to a given series of questions, the computer should for all intents and purposes be treated as human. Now known as the Turing Test, this idea of the imitation game has become embedded in popular culture due to a scene in the 1982 film *Blade Runner* where a detective asks the person opposite him a series of questions and, based on his answers, evaluates whether he is a human or an android.

The paper in *Mind* demonstrates Turing's long-term interest in the following question: If "dumb" electrical circuits in a computer could perform mathematical tasks previously only carried out by human minds, was it possible that similar "dumb" processes ultimately underpinned all the ways those minds worked? Though, of course, the components in a brain's circuit would be interacting chemicals in nerve cells rather than electrical valves and relays.

To answer this directly, Turing knew, would be well-nigh impossible. Even if the brain is a circuit of simple chemical interactions, it's a circuit with billions of such components. So, as a first step, Turing decided to investigate a simplified version of another biological process to see if it could be explained by the actions of a simple chemical "circuit." His goal was to demonstrate that complex biological behavior could derive from, what deep down, are simple processes.

This was his motivation for one of the most ambitious papers he ever wrote, "The Chemical Basis of Morphogenesis." Submitted in late 1951, this paper was nothing less than an attempt to propose a mechanism by which embryos are shaped as they develop in the womb. Turing considered it his best work since his 1936 paper laying down the foundations of computing. By any measure it's a work of extraordinary scientific imagination. For in it, Turing completely recast the second law of thermodynamics. Here it is perhaps important to remember that since the discovery in the mid-nineteenth century that entropy always increases, the second law has often had strongly negative connotations. The inevitable dissipation of energy, such as the flow of heat from hot to cold, became seen as synonymous with decay and death. Dissipation, the smoothing out of all variation in the universe, was seen as the reason beautiful and intricate systems such as living creatures degrade and die.

Turing turned these pessimistic associations on their head, arguing that dissipation didn't solely cause decay, but could *create* structure and form. Under certain conditions, he suggested, as certain substances diffuse and spread out, they *self-organize* into patterned structures. These pattern-creating substances he named morphogens, arguing that as they diffuse through the cells of an embryo, they also shape that embryo.

Put another way, Turing was trying to explain how embryos, which start off as a single cell—the fertilized egg known as the zygote—could divide into multiple, essentially identical cells, which later differentiate into the specialist cells that, arranged in highly organized ways, make up a living organism. Look at your hands, for instance. Given that every one of the small number of identical cells you once were contained a complete set of your genes, how did the cells that formed your hands know only to turn on the genes relating to hand formation? Why didn't they instead produce a foot at the end of your arms? Turing felt that the key to understanding this biological tailoring lay in the diffusion of morphogens. This process, he wrote, is "a possible mechanism by which the genes of a zygote may determine the anatomical structure of the resulting organism."

The idea that diffusion can create structure is counterintuitive. As two modern developmental biologists, Jeremy Green and James Sharpe, put it, "Just think of a drop of ink in water—diffusion will cause the slow but sure dispersal of the ink molecules until all the water is faintly colored. The original pattern—a spot—is destroyed; the final state has no spatial heterogeneities and thus no pattern. Diffusion would seem to be a quintessentially entropy-increasing, disorder-maximizing process. The idea that diffusion itself could create a pattern—that it could drive the well-mixed ink back into a spot—was (and still is) very surprising."

Another concept is at the heart of Turing's paper. Most likely inspired by his experience with electrical circuitry during the war, this is the phenomenon that engineers call feedback. There are two kinds, positive and negative. A famous (post-Turing) example of positive feedback is the howling noise emitted when you hold a powered-up electric guitar close to its loudspeaker. This is started by a tiny, inaudible vibration in a guitar string sending a small electrical signal to the amplifier, which turns it into a quiet but audible sound. Like all sound, this is an oscillating wave of air pressure, and it causes the original guitar string to vibrate in sympathy but by a greater amount than it was before. This in turn sends a larger electrical signal to the amplifier, causing a much louder sound to come out the speaker. This makes the guitar string vibrate even more violently, which then causes an even bigger electrical signal to go to the amplifier, and so on in repeating loops. In a short time, we hear a deafening howl.

Though Turing never heard electric-guitar howl-round of the kind made famous by Jimi Hendrix, Turing's wartime work had included designing radio communications systems prone to this kind of positive feedback. So he would have known that positive feedback occurs when a cause has an effect that loops back and *increases* what caused it the first place. Negative feedback, by contrast, occurs when a cause has an effect that *diminishes* what caused it. A good example is a domestic heating system that's controlled by a thermostat. When heat from the radiators causes the temperature to exceed a certain level, the thermostat switches off the boiler, and the temperature falls below that level. At that point, the thermostat switches the boiler back on, and the temperature goes back up, and over time the room's temperature remains at a near-constant level. As a rule, positive feedback causes systems to spin out of control, whereas negative feedback keeps them stable.

Turing argued that feedback, both positive and negative, can occur when certain chemicals react, and that morphogens are examples of such chemicals. He then argued that when chemicals of this kind diffuse through a body of identical cells, they can trigger changes in them that, as the cells differentiate, cause them to form a recognizable pattern.

An example of the kind of pattern Turing had in mind were the stripes on a zebra's skin. The skin cells of the zebra start off identical. But as morphogens capable of feedback diffuse through them, they cause some cells to turn dark and others to turn white, such that, when observed from a distance, a striped pattern emerges. Turing did not provide the chemical formulas for such morphogens, but focused instead on proving mathematically that they could, in the right circumstances, spontaneously create patterns out of nowhere. "Such a system," wrote Turing, "although it may originally be quite homogeneous, may later develop a pattern or structure."

Turing then provided an important caveat. The chemical processes that occur in a real living creature, he knew, would be far more complicated than anything his mathematics could describe. His aim was simply to demonstrate that diffusing morphogens could create structures and thus establish the principle behind spontaneous pattern formation. As he put it, "This model will be a simplification and an idealization."

Turing pictures a hypothetical arrangement of cells that is indeed highly simplified—a ring of identical cells. He then describes circumstances under which, as two morphogens flow across these cells, a pattern will emerge. Some turn black and others white. They do so in a regular, patterned way. So, for example, if the total number of cells in the ring is one hundred, there will be ten black ones, followed by ten white ones, followed by ten black, and so on. From a distance it will appear that a ring has changed from being the same throughout to having a striped pattern.

This can happen, Turing shows, if the two morphogens have specific properties. The first morphogen, let's call it X, turns a cell black. The second, Y, turns a cell white. X must also be capable of positive feedback. That means that when one molecule of X reacts with another molecule of X, it creates a third molecule of X. In other words, X generates more copies of itself as long as there is a ready supply of free energy and the raw materials from which X is made. Y, on the other hand, acts to create a negative feedback loop that can reduce the production of X. This means that if the levels of X exceed a certain amount, a molecule of Y destroys a molecule of X, thus stopping the production of more X. Y, if you like, acts as a thermostat that regulates the production of X.

In unpublished notes, Turing explained how this system might create patterns. To do so, he imagined a circular island whose inhabitants live only along its circumference—on the seashore. These people, who move at random along the shore, much like diffusing morphogens, are either cannibals or missionaries. Both kinds can die, but the cannibals can reproduce with one another, and so like morphogen X, they can increase in number. The missionaries, being celibate, cannot reproduce. But if two missionaries encounter a cannibal, they convert him or her into a missionary. This increases the number of missionaries but puts a brake on the rate at which the cannibal population increases. The missionaries are like morphogen Y, which can brake the rate at which morphogen X increases.

Turing then analyses mathematically how such a system will evolve over time. If the number of cannibals far outweighs the number of missionaries, then the missionaries will soon die out and the island's entire shore will be populated with cannibals. If the number of missionaries far outweighs the number of cannibals, then the cannibals will soon all be converted, and the island will only be populated by missionaries.

But under certain circumstances, if the proportion of missionaries to cannibals falls within a certain range, and the rate at which the two groups can move around the island falls with certain values, then a stable pattern of missionaries and cannibals emerges. There will be cannibal zones alternating with missionary zones all around the shore of the island, and all the zones will be the same size. Similarly, in the ring of cells, if the proportion of X to Y is right, and the rates at which they can diffuse are also right, then stable, equal-sized zones—where X dominates in one and another in which Y dominates—will appear around the ring of cells. And because X turns a cell black and Y turns it white, the ring will develop a striped pattern.

The devil in this argument lies in solving the mathematical equations that describe this situation so that one can predict what proportion of X to Y, and what diffusion rates of the two morphogens involved, will lead to stable patterns. And as Turing points out, a circle of cells is not representative of the way that cells are arranged in living creatures. Analyzing cells clumped together in more realistic three-dimensional shapes makes the mathematics even harder. Solving such equations is so difficult, he writes, "that one cannot hope to have any very embracing theory of such processes, beyond the statement of the equations."

That means finding the optimal "pattern-inducing" solutions to the equations by trial and error. One must try innumerable combinations of the ratios of X to Y and their respective diffusion rates to find the few that work. This is tedious to do by hand. Yet, Turing points out, it is a task for which computers are ideally suited. So, as the Manchester University machines grew more powerful, Turing started writing programs that searched for solutions to the equations describing morphogen diffusion so that he could investigate the patterns that might form. "Our new machine is to start arriving on Monday. I am hoping as one of the first jobs to do something about 'chemical embryology,'" he wrote with great excitement to a friend in 1951.

By modern standards the Manchester machine of that time was glacially slow and cumbersome. Nonetheless, with its help and with "a few hours by a manual computation," Turing was able to draw a picture that

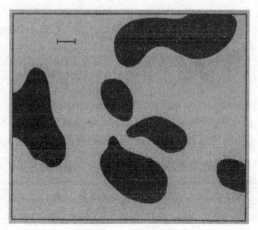

An example of a "dappled" pattern, as shown
in Turing's paper

showed how a system of diffusing morphogens could produce a dappled black-and-white pattern of the kind seen on the hides of cows.

This picture is the first to show that a mathematical process can produce a plausible biological shape. Turing's work marked the birth of a new field of science. Today the use of computers to model these kinds of processes is ubiquitous. It's important to stress, as Turing does in his paper, that this kind of pattern formation fits naturally within the laws of thermodynamics. For such structures to be formed, he writes, "a continual supply of free energy is required." His aim was to show how free energy, which nearly all life on earth obtains directly or indirectly from the sun, drives the creation of pattern and structure as it dissipates.

Another way to understand this kind of dissipative pattern formation is to consider the way arrays of sand dunes form in the desert or on long beaches. Initially, a strong, steady wind blows across a featureless flat sandy beach. This wind is, in effect, a steady supply of free energy. But as it blows the sand along, there are initially no obvious reasons for dunes to form. Beaches are, however, never perfectly smooth. An object such as a small rock or log is usually on it somewhere. Wind pressure will then cause sand to build up on the windward side of the rock, and this induces a positive feedback loop. As more sand builds up, a greater obstacle to the wind is created, which causes more sand to be blown into the obstacle, making it even bigger, and so on.

Then negative feedback comes into play in the form of gravity. When the dune reaches a certain height, the sand can no longer be supported, and it starts to cascade down the leeward side of the dune. By this time, however, the dune is high enough to create a wind shelter. So the next dune must form some distance away, where the first dune no longer provides shelter from the wind. If the wind speed and the size and the stickiness of the sand fall within appropriate parameters, an array of sand dunes forms. What started as a tiny inconsistency such as a small rock has turned into a beachwide pattern. Turing's genius was to see that a similar process, via chemical reactions, might shape embryos.

Turing had given scientists a bold and unprecedented means of understanding how living shapes are formed. But it was only a first step, which he knew needed a great deal of further thought. Tragically, however, Turing would die long before he could see where these ideas led.

Turing sent his paper on morphogenesis to Britain's preeminent scientific body, the Royal Society, in the fall of 1951. He was justifiably proud of this achievement and seems to have regarded it as a significant breakthrough. We know this because three years later, in May 1954, Turing wrote the beginnings of a short story about a fictional scientist called Alec Pryce. There's little doubt from the few pages of the story that survive that Pryce is Turing. Like Turing he is homosexual, like Turing he has given talks on the BBC, and like Turing he had published a significant paper while still in his twenties. In what is clearly a reference to the morphogenesis paper, Turing writes, "This last paper was real good stuff, better than he'd done since his midtwenties." Then comes the admission that this achievement inadvertently triggered the first in a series of events that led to tragic consequences. The story describes Alec as being open about his homosexuality and goes on to say, "It was quite some time since he had 'had' anyone, in fact not since he had met that soldier in Paris last summer. Now that his paper was finished he might justifiably consider that he had earned another gay man, and he knew where he might find one who might be suitable."

The Alec Pryce story suggests that Turing, on completing the intense intellectual effort he'd put into his morphogenesis paper, decided to seek a sexual partner. He succeeded. In December 1951, a few weeks after the paper's completion, he did indeed meet a young man, Arnold Murray, in a seedy part of Manchester. Over the next few weeks, they had sex several times in Turing's house. Then events took a sinister turn. An acquaintance of Murray's, on hearing of the affair, burgled Turing's house. To the burglar, a university academic appeared a well-paid and inviting target. He also reasoned that a gay man who lived on his own could, at a time when homosexuality was illegal, be blackmailed against going to the police with the threat of revealing his sexuality. The property stolen, amounting to around £50, included a compass, some clothes, a fish knife, and a watch that Turing had been given by his father.

After the break-in, Turing confronted Murray. Although Murray had taken no part in the robbery, he confessed that he had identified Turing as a possible target to the burglar. Furious, particularly over the loss of his father's watch, Turing reported the incident at his local police station.

This was a disastrous decision. The police concluded after talking to Murray that he and Turing were in a sexual relationship and this was worthy of further investigation. On February 11, 1952, the police interrogated Turing. Once again naively faithful to the truth, Turing confessed and was immediately charged with gross indecency, the same crime for which Oscar Wilde had been convicted in 1895. On March 31, 1952, Turing pleaded guilty in court to the charges. His sentence was that he could avoid going to jail if he agreed to something called organotherapy. This meant being given high doses of a drug, a synthetic form of estrogen, designed to reduce or remove the male libido. Turing agreed and for a year he took the drug, first in pill form and then as an implant in his thigh. This turned off the production of testosterone in his body, effectively castrating him. Turing showed two symptoms common in men whose testicles malfunction—he developed breasts and found it hard to concentrate, writing, "I've got a shocking tendency at the moment to fritter my time away in anything but what I ought to be doing." Experts on the male reproductive system now know that drugs of this kind cause men to put on weight, lose body hair, and become impotent and extremely lethargic.

By accepting the treatment, Turing was able to keep his job at Manchester University and continue to have access to the computer there. The many pages of handwritten notes later found in Turing's home make it clear that in the years following his sentence, while battling the effects of stilbestrol, he worked extensively to develop his theory of morphogenesis. This material shows that he was making some progress toward formulating a general mathematical understanding of phyllotaxis, the way in which plants arrange leaves around their stems. Most of the other writings are hard to judge. Turing's handwriting was terrible, and though some of the mathematical formulas he's playing with are legible, overall, they've proved to be indecipherable. His friend and fellow mathematician Robin Gandy later said of them, "It will be difficult, in some places impossible, to know exactly what the fragments are about."

We shall never know if Turing was on the brink of a new breakthrough. For on June 8, 1954, his housekeeper discovered him dead in bed at his Manchester home. The autopsy decided the cause was cyanide poisoning, with four ounces of cyanide fluid found in his stomach. The inquest was brief, and the coroner quickly concluded that Turing's death was "a deliberate act." The verdict was suicide "while the balance of his mind was disturbed."

Not everyone accepted the verdict. Turing's mother, for one, preferred to believe her son's death an accident, as at first did his brother, John. Turing made a habit of carrying out chemistry experiments at home and had set up his own apparatus to electroplate cutlery with gold, a process that requires cyanide. The previous Christmas, Sara Turing had warned her son of the dangers. Colleagues at Manchester University, along with Turing's neighbor, supported Sara's view that the death was an accident, remarking that he had shown no signs of distress in the preceding weeks and days. That he left no note adds further ambiguity.

In truth the coroner was probably right. It's hard to drink so much cyanide by accident. There is also evidence that, despite outward appearances, Turing had been in mental distress in the two years following his arrest and conviction. Additionally, he was aware that the authorities were keeping close tabs on him because he had a criminal record and because of his recent top-secret code-breaking work. Turing wrote to a friend a year after his sentencing. "If I had so much as parked my bicycle on the wrong side of the road there might have been 12 years for me. Of course the police are going to be a bit more nosy, so virtue must continue to shine." The stress of surveillance and of constantly being on good behavior must have taken its toll. So much so that Turing began to see a psychoanalyst in Manchester, Dr. Franz Greenbaum, shortly after writing this letter.

Greenbaum and Turing got on well. Greenbaum held no negative views of homosexuality and treated him as a friend as much as a patient, inviting Turing to his home on several occasions. And it's from Greenbaum that we have perhaps the most compelling evidence that Turing took his own life. On Greenbaum's encouragement, Turing wrote three volumes of dream diaries during his year of psychoanalysis. Shortly after his death, Turing's older brother, John, read through them. They changed his mind. According to John's son, Dermot Turing, "It leaves John in no doubt that it's not an accident and that Alan had killed himself." John was appalled by what he read in the diaries, which apart from revealing much about Alan Turing's state of mind, contained "all this stuff about his hatred of his mother and frankly it paints a picture which is deeply unwelcome." To spare his mother further distress, John decided that the diaries had to be destroyed, which he duly did.

Dermot Turing makes another telling point about his uncle: as well as feeling persecuted and watched, Alan was lonely. He was not close to his family, and harboring secrets both personal and professional, he could not talk to anyone with complete candor. Though his brother and mother had found out about his sexuality after his arrest, neither felt comfortable discussing it. Dr. Greenbaum was the one person with whom Turing could talk openly about personal matters, but because of the Official Secrets Act, he could never discuss many aspects of his work and life during the war. There is no way of proving categorically that Alan Turing committed suicide. But that doesn't diminish the tragedy of his death or the injustice he suffered at the hands of the society and nation for which he'd done so much.

Turing's death at forty-one was an incalculable loss. Had he lived another two or three decades, he would have witnessed an exponential growth in computing power, which might well have nurtured his ideas about morphogenesis and embryo formation. In any event, his later work lapsed into obscurity, not least because physicists and mathematicians of the 1950s regarded biology as a discipline far removed from their own. Worse, biologists found the mathematics in his papers daunting.

Turing's concept of spontaneous pattern formation through dissipation was dealt a further blow in the late 1960s and early 1970s when a conceptually simpler explanation for the way embryos are shaped emerged. This idea became known as PI, standing for *positional information*, the champion of which was a developmental biologist named Lewis Wolpert. Born in South Africa in 1929, Wolpert, after first training as a civil engineer, switched to biology at King's College London, where he completed a PhD in the mechanics of cell division. PI, unlike Turing's theory, requires no complex mathematics. It hypothesizes that morphogens of different kinds exist in varying concentrations in different parts of the embryo. The concentration of a particular morphogen at a specific point causes a cell to develop one way or another.

A popular explanation is the so-called French flag model. Imagine a rectangular array of initially identical cells in a bath of morphogen solution. The concentration of the morphogen drops from left to right. In other words, in the left third of the rectangle, working from left to right, it drops, say, from 100 to 70 percent concentration; in the middle third it falls further, from 70 to 30 percent; and in the right third, it drops from 30 percent down to 0. Now, let's say a high concentration of morphogen causes cells to turn blue, whereas a moderate concentration causes them to turn white, and a dilute concentration cause them to turn red. It's not hard to see that the result would be a color pattern resembling the French flag—three bands, from left to right, that are blue, white, and red.

From the beginning, Wolpert regarded the PI model as conflicting with

Turing's ideas, writing in 1971 that they were "the antithesis of positional information." Indeed, the experimental data appeared to support Wolpert's theory. For in the late 1980s, the Nobel Prize–winning biologist based at the University of Tübingen in Germany, Christiane Nüsslein-Volhard, identified, with her colleagues, a morphogen that plays an important role in shaping the larvae of fruit flies. This chemical, named biocid, was the first morphogenic, shape-imparting chemical to be isolated, and it seemed to work according to Wolpert's PI theory, rather than Turing's dissipative pattern-forming explanation.

Fruit fly larvae resemble tiny worms that are about ten millimeters long. Their cylindrical bodies are made up of a row of eleven segments, each just under a millimeter in length.

In the late 1980s and early 1990s, scientists studied how morphogens such as biocid determined the dimensions of the fruit fly larva's segments. They found compelling evidence that they resulted from varying concentrations of morphogens—as suggested by the PI theory. Further revelations in the last decade of the twentieth century seemed to cement it as the explanation for morphogenesis over Turing's theory.

But as the first years of the twenty-first century unfolded, evidence mounted that spontaneous pattern formation driven by diffusion, as described by Turing's morphogenesis paper, does indeed occur in the living world. Initially, scientists discovered evidence in patterns that occur across a species but where each individual member of that species manifests a unique version of that pattern. Take for example the distribution of hair follicles in mammals, humans included. All of us have a two-dimensional array of hair follicles on our heads, but the individual location of each follicle varies from person to person. Such patterns are hard

Fruit fly larvae

to explain with PI because it implies that each embryo starts off with a unique pattern of morphogen concentrations, which then gave rise to that individual's unique pattern of hair follicles. That raises the question of what created the pattern of morphogen concentrations in the first place. Turing's theory, by contrast, easily enables the formation of similar but nonidentical patterns across individual members of a species.

Remember, as with the formation of sand dunes, that a tiny initial difference is all that's needed to kick-start pattern formation. This can be caused by the random wobbling of molecules that occurs all the time or indeed by a nonrandom trigger that's coded in the genes. Turing's equations predict that although the patterns created in this way will be remarkably similar, they will never be identical. That's because the tiny "jiggles" that start the process are themselves never identical, and so the pattern they eventually produce is never identical either. Imagine taking a picture of the same beach at the same time, year after year. Each picture will show similar arrangements of sand dunes, but no two pictures will be identical because the imperfection that starts the process from year to year will be different.

By studying mice, a team of researchers in Germany found compelling evidence that two morphogens, proteins named WNT and DKK—WNT being the positive-feedback "cannibal" morphogen and DKK being the negative-feedback "missionary" morphogen—were responsible for the way that hair follicles are arranged. Researchers in Japan also argued convincingly that the striped patterns in angelfish and tiny zebra fish are created by a Turing mechanism.

Then in 2012, around the hundredth anniversary of Turing's birth, a flurry of further papers appeared that seemed to back his theory. One, by Professor Jeremy Green and his team of developmental biologists at King's College London, provided the most compelling evidence yet.

That team's initial interest lay with studying how the face forms in the womb, with a particular focus on the formation of cleft palates and other abnormalities. To do so, they investigated the formation of the ridges, known as rugae, that form during gestation in the upper palate. If you rub your tongue along the upper palate, you can feel them—humans have four, mice have eight.

The scientists identified the two morphogens—cannibal and missionary—that were creating the pattern. The first was a chemical called fibroblast growth factor, or FGF for short, and the second a chemical called Sonic hedgehog, or Shh. By altering the amount of the two morphogens in mice embryos, the research team discovered that they caused changes in the number of rugae that were formed in their mouths, exactly as Turing's equations predict.

This rugae paper was followed two years later by another extensive investigation from a team led by Professor James Sharpe at the Centre for Genomic Regulation, based in Barcelona. They uncovered how morphogens shape our hands in accordance with Turing's predictions. Hands or paws in all vertebrates can be thought of as a striped pattern. After all, our hands consist of five roughly parallel fingers or digits that repeat. Using a combination of computer modeling and observations of mouse embryos, the team uncovered the morphogens involved in creating these patterns of digits. Three chemicals, Sox9, BMP, and WNT, each play a role in this process. Although this is a somewhat more complicated arrangement than Turing's two-morphogen "cannibal-missionary" model, it works in a similar way. The proof that this "digit" formation in mice was an example of a Turing pattern came when the team varied the relative amounts of the three chemicals in the mice embryo. Their computer models predicted that a certain ratio of the chemicals would, for instance, cause a mouse to develop three fat digits rather than five—and that is indeed what they observed.

It seems living organisms use a combination of Turing-style spontaneous pattern formation and Wolpert-style PI to create the myriad shapes that we see in the living world. In regard to our hands once again, biologists believe that Turing mechanisms create a template of five digits, and a PI-style morphogen gradient gives each individual finger its distinctive shape. In other words, our hands have five fingers thanks to a Turing system, but our thumbs, forefingers, middle fingers, ring fingers, and pinkies look different from one another because of PI. It's a mark of how strongly Turing's ideas on embryo formation have returned to become a centerpiece of developmental biology that shortly after the paper on digit formation was published, Lewis Wolpert, originally a harsh critic, gave an interview in which he acknowledged their validity and described Turing as a "genius."

The science of embryo development is still itself at an embryonic stage. Nonetheless, we are moving closer to understanding how heart valves, lungs, and more develop. In decades to come, these insights may lead to treatments for illnesses and birth defects that we can today only fantasize about.

Some critics claim that scientists, in their desire to explain everything, reduce the wonders of the universe to little more than equations and chemical reactions. To which I say, stand on a beach one day and look at the waves and the patterns of sand dunes through the fingers of your hand. Consider that all these phenomena are connected by the same underlying principles of nature. Consider that all these beautiful patterns emerge from the dissipation of free energy and all start as tiny imperfections.

# CHAPTER NINETEEN

# Event Horizon

Bekenstein and Hawking were the first to enter a remote country and bring back gold.

—Theoretical physicist Leonard Susskind

Your idea is so crazy that it might just be right.

—The physicist John Wheeler to this then student Jacob Bekenstein

Thermodynamics had come a long way by the 1970s. Its laws had become foundational to the pursuit of biology, chemistry, engineering, and physics. But one area of science held out: in the far reaches of space, there were believed to be phenomena that, alone in the universe, did not behave according to the principles of thermodynamics. Specifically, they appeared to defy the second law, namely that the entropy of a closed system such as our universe always increases. These objects, the most outlandish prediction of Einstein's theory of general relativity, are black holes.

Black holes are bizarre regions of space into which anything can fall but from which (almost) nothing can escape.

These strange entities are a consequence of Albert Einstein's magnum opus, the general theory of relativity, which he published in November 1915. It aimed to extend the ideas of special relativity, which had described the consequences of assuming the laws of physics are the same for all observers, irrespective of the speed at which they're moving. Special relativity did not, however, investigate the consequences of that same assumption if the observers' speeds are varying. How can one create consistent physical laws for all observers, including those who are accelerating and, in particular, moving under the influence of gravity? This was the question the general theory of relativity sought to answer.

Einstein understood this meant replacing Isaac Newton's theory of gravity, which had been published in 1687. It also meant that the redefinition of space and time brought about by the special theory had to be made even stranger. For an intuitive grasp of this new vision of reality, we might re-create in our minds one of Einstein's most famous thought experiments, first imagined in 1907, which he later described as the "happiest thought of my life."

Imagine a physicist called Alice in a windowless box in the depths of empty space, far from the gravitational effects of stars and planets. She floats around the box, feeling no pull in any direction. When Alice tries to weigh herself by strapping some scales to the underside of her feet, she exerts no downward pressure on the scales. Her weight therefore registers on the scales as zero—she is weightless.

Now, imagine that instead of the box being in the depths of space, it is fifty kilometers above the earth's surface, directly above Greenwich in London at longitude 0°. The box, we discover, is plummeting downward in free fall. That means the box and its inhabitant, Alice, accelerate at greater and greater speeds toward the earth's surface. But importantly, they are accelerating downward at precisely the same rate. This idea, that objects fall at the same rate irrespective of their mass, had been known since the time of Galileo. Indeed, Vincenzo Viviani, one of Galileo's assistants, recalled the great physicist demonstrating this property by dropping two objects of different mass from the top of the Leaning Tower of Pisa and showing that they hit the ground simultaneously. Historians are unsure how true the story is, but it is certainly the case that Galileo had rolled balls of different masses down a ramp and showed that they all moved a given distance in the same time.

Now let's return to Alice in her falling box above Greenwich. Because Alice and the box accelerate downward at the same rate, she will feel that she is floating around in the box in the same way that she floated around in it when it was in empty space. If Alice weighs herself, the scales will again register her weight as zero. For all intents and purposes, Alice will have no way of knowing that she is hurtling toward the ground at ever-greater speeds until the box smashes into the earth's surface. Until that fatal moment of collision with the ground, the earth's gravity is undetectable from inside the box. As Einstein put it, "For an observer in free fall . . . there exists, during his fall, no gravitational field." This observation, that being in free fall is indistinguishable from being in a region of zero gravity, is called the principle of equivalence.

Now extend this picture. Imagine a second box, also fifty kilometers above the earth's surface and also above the Greenwich meridian at a longitude of $0°$. But this box is fifty kilometers due south of Alice and falling at the same time as her. This second box is made of a heavier material than the first one. Inside is another physicist, Bob, who weighs twenty kilograms more than Alice.

Watching these two boxes plummet downward is Cleo, a physicist looking up from the earth's surface, on which she stands. Cleo has long-distance X-ray vision that allows her to see through walls. But what does she see? She sees that everything, the boxes and their respective passengers, all increase their speed at the same rate as they accelerate downward. In addition, Cleo will observe that the two boxes are not only falling, but they are also moving nearer and nearer to each other; the horizontal distance between them is decreasing. They do so at a much slower rate than the rate at which the boxes fall, but they do steadily approach each other.

That Cleo sees the two boxes moving toward each other is intriguing. To explain that, we should turn first to Isaac Newton's law of gravity. This asserts that a force is pulling everything toward the earth's center. Because the force acts inward to the same point, the earth's center, while pulling the two boxes downward, it draws them closer to each other as they fall.

This Newtonian force explains what Cleo sees, but it begs many questions. First, all of the various falling objects—the two people and the two boxes—are of different masses. Nonetheless, they all accelerate downward by the same amount, implying that the force pulling each object not only varies, but does so by exactly the right amount to ensure their same rate of acceleration. But how does the earth do that? How does it vary its pull precisely to account for the mass of each object that it pulls? Then there's the radial nature of the earth's pull. It's as if there's a mysterious "line-of-sight" communication between the earth's center and every object in its vicinity. The earth, in other words, appears to measure the distance from its own center to every object nearby, as well as measuring the mass of all these objects. Then it has to calculate the exact amount of force to exert on each of those objects and the direction in which those forces must act. It then instantly transmits that force to all objects falling downward.

It's a preposterous idea. Indeed, the first person to point out quite how preposterous was Newton himself. In a letter shortly after he published his theory of gravity, he said, "That gravity should be innate, inherent, and essential to matter, so that one body may act upon another at a distance through a vacuum, without the mediation of anything else . . . is to me so great an absurdity that I believe no man who has in philosophical matters a competent faculty of thinking can ever fall into it. Gravity must be caused by an agent acting constantly according to certain laws; but whether this agent be material or immaterial I have left to the consideration of my reader." It's worth noting that no other force in nature acts like gravity. For example, a paper clip and a heavier object such as a bolt will accelerate toward a magnet at different rates. But if you drop the paper clip and the bolt, they will fall downward with exactly the same acceleration.

The general theory of relativity is Einstein's considered response to Newton. It dispenses with the absurd notion that objects such as the earth perform complex calculations and act on other objects instantly at a distance through empty space. According to Einstein, something very different happens when things fall. A massive body such as the earth has no sense of the objects in its vicinity and exerts no force on anything at all. Instead, the presence of a mass such as the earth curves the space around it and slows down the passage of time in its vicinity.

This idea, that the void itself can curve and that gravity can affect time, is perhaps the most radical in all of science. To appreciate it requires jettisoning many of our commonsense notions. The arguments are mathematically so complex that Einstein took eight years to formulate them. His "happiest thought" that a person in free fall feels no gravity occurred in 1907, and the general theory was published in November of 1915.

Let's return to Alice inside her sealed box when it's floating in empty space. To all intents and purposes, Alice is stationary in space, but she is moving forward in time. To visualize this as Einstein did, imagine a graph in which the vertical axis is time and the horizontal axis is space. For simplicity's sake, let's assume that there's only one spatial dimension, left or right, along the horizontal axis. In this representation, Alice isn't stationary. She is moving upward along a straight line parallel to the vertical "time" axis. That simply means that her spatial position doesn't change but she is moving into the future.

How does this compare with what Alice experiences when she's in freefall toward the earth? Remember the principle of equivalence, which states that from Alice's point of view nothing has changed; she continues to believe that she and the box are all moving in a straight line through time and space. (The box Alice inhabits has no windows, so she can't see the earth becoming nearer.)

So how is it that Cleo sees Alice speeding up and drifting closer to Bob, who's falling in the second box? The answer is that the mass of the earth curves the space through which they're moving in a way similar to that shown in the picture below.

The earth's mass curves what was flat space toward the point N. That means that Alice, though she thinks she's moving forward in time and not moving in space, is in fact following the curved line from point A to point N. And though Bob, like Alice, thinks he's moving forward in time and not moving in space, he, too, is following a curved line, from point B to N. So, Alice and Bob move inexorably closer to each other and to point N not

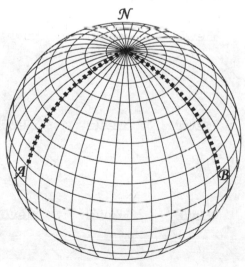

A massive object curves flat space and time
so that Alice's and Bob's "straight-line"
paths meet at N.

because they're being pulled by a gravitational force, but because that's the path they must follow in curved space if they remain still.

So the key idea to general relativity is that Newtonian gravity is an illusion. We think the earth is pulling us down with a force. It isn't. Its mass has curved space in such a way that a straight line in this curved space leads toward the earth's center.

Einstein presented the mathematical equations of the general theory of relativity at a meeting of the Prussian Academy of Sciences in Berlin in November 1915. In the years that followed, the theory was triumphantly vindicated by observations of the cosmos. Because space near a massive object such as our sun is curved, the theory predicts that a light beam from a distant star traveling past the sun will follow a curved path. And that is indeed what happens. As the picture below shows, light from a distant star bends as it passes the sun. To an observer on earth it looks as if the star has shifted in position because we assume the light traveled in a straight line, when it didn't.

This effect can be discerned during a solar eclipse, when the stars behind

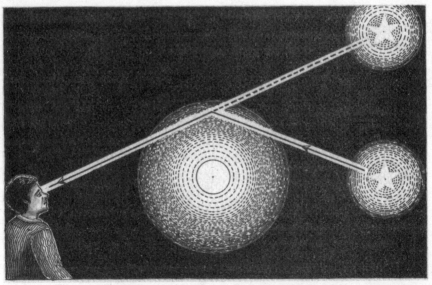

An exaggerated diagram showing gravity can make a star's position
appear to shift.

the sun become visible. During the day, the sun's brightness obscures the light from stars that are behind it. During an eclipse, however, when the sun is blocked by the moon, day becomes night, and for a few moments those stars can be seen. That was why in 1919, just four years after Einstein's presentation of his theories, a British team of scientists traveled to Brazil and West Africa to photograph the positions of stars behind the sun during an eclipse. When they compared those observations with the stars' positions six months later, they saw a small difference. This proved that the stars' light had indeed bent when it passed by the sun six months before, but not later when the sun was no longer in its path. Many observations have subsequently been made of the way that massive objects cause light beams to follow curved paths.

Yet even though the scientific community rapidly accepted general relativity, only a few scientists studied it for the first half of the twentieth century, in part because it represented an immense rethink and also because its mathematical formulas were fiendishly difficult to follow. The predictions that these made, in the main, were rather close to those predicted by the much-simpler rules of Newtonian gravity. The latter may have been philosophically "absurd," but they were much easier to work with.

Interest finally grew in the years following World War II, when physicists became interested in the behavior of extremely large objects that were many times heavier than our sun. With these, general relativity made very different predictions to those of Newtonian gravity. Investigating such objects therefore promised new insights into hitherto unexplored ways in which the universe might work. Of great interest was an aspect of general relativity that had been spotted only a few weeks after Einstein first announced his discovery.

In early 1916, Karl Schwarzschild, a German physicist and astronomer then serving on the Russian front, had published an analysis of Einstein's theory. This work made the disturbing prediction that if a massive object such as a star was compressed to a sufficiently high density, it would curve space and slow down time to such an extent that something bizarre would happen. Space and time would become infinitely curved. This supermassive star would create a "singularity," which is a fancy way of saying that the mathematics of general relativity fail and can make no description of what might be going on.

A space-time singularity

Many physicists, including Einstein, had dismissed such singularities, claiming that they were unlikely to exist in the real universe. One after another, however, all the reasons people had put up for dismissing their reality fell away, and by the late sixties and early seventies, some of the world's brightest physicists focused their minds on these singularities to see what secrets they might contain. No such objects had yet been detected in space, so all such work remained theoretical. Yet despite the lack of proof, these singularities—or gravitationally collapsed objects, as they were also known—earned an evocative nickname: *black holes*.

Imagine a shallow ocean of water that stretches to an infinite distance in all directions. The only living inhabitants of this ocean are a species of blind fish who swim in its water but can't feel it. So, if the water surrounding one of the fish is flowing, the fish moves along with the water but without any idea that it's doing so. To communicate, the fish, who have exceptionally good hearing, send sound signals at a fixed speed through the water. One key aspect of this water world is that the speed at which sound travels is the fastest possible speed at which anything can move.

In one part of the ocean, a drain hole sucks in all the water around it. The closer the water is to the drain hole, the faster it moves. At a specific distance from the drain hole, which is determined by how powerfully it sucks the water, *the inward speed of the water flow reaches and then exceeds that of the speed of sound.*

It helps to visualize a circle around the drain hole. Outside it, the water flows more slowly than the speed of sound. Inside it, the water flows toward the drain hole more quickly than the speed of sound. On the rim of the circle, the water flows at exactly the speed of sound. Let's call this circle the sonic boundary.

Now imagine two blind fish. One, Alice, is far enough away from the drain hole for the flow of water to be unaffected, whereas the other, Bob, is much closer, so much so that he is moving along with the water toward the hole. But because he's moving with the water, he's unaware of this motion. To stay in contact, Bob and Alice have agreed that once every second he will make a sharp sound that sounds like a ping.

Initially Alice hears a ping from Bob every second. But the intervals between each ping become longer and longer. That's because, unbeknownst to her and Bob, the water around Bob is moving toward the drain hole and away from Alice. Bob's pings have therefore to move against the flow of the water to Alice, and they take a longer time to cover the distance from Bob to Alice than they would if the water were still. The closer Bob moves to the sonic boundary, the more pronounced this effect becomes, and eventually the intervals between pings become so long that Alice thinks Bob has stopped signaling her. Alice surmises that Bob has been moving away from her more and more slowly, and that his watch seems to tick more and more slowly as it nears the sonic boundary. At the boundary, it seems to Alice that Bob's watch has stopped.

To Bob, however, the experience feels very different. As far as he's concerned, he's been dutifully sending pings every second, and he has no awareness that he's been carried toward the drain hole or crossed the sonic boundary. All seems well until he decides to turn around and return to

Alice. Then he finds that however hard he tries, he cannot go back over the sonic boundary to the part of the ocean he came from and where Alice remains. That's because the water on Bob's side of the sonic boundary is flowing faster than the speed of sound toward the center of the drain hole. Because nothing in the water world can travel faster than this speed, try as he might, Bob cannot escape back to the region of water from which he came. In fact, however hard he swims, he will inexorably be swept into the center of the drain hole.

So what's the relevance of this drain hole to a black hole? The analogy works, roughly, as follows: The drain hole sucking water toward it is equivalent to the singularity at the center of a black hole sucking space toward it. Just as water starts to flow faster than the speed of sound at a circle around the center of the drain hole, so too, there is a spherical surface around the singularity at the center of a black hole at which *the speed of the flow of space exceeds the speed of light*—yes, think here of empty space as you would a flowing liquid. Because no objects or signals in our universe can move through space faster than the speed of light, everything within this spherical surface is doomed to stay within it. Just as Bob could not go back across the sonic boundary because he could not swim faster than the speed at which the water is flowing inward, so, too, anything within the spherical surface around the singularity cannot go back across it. And just as Bob, the fish in the water world, is doomed to be swept by the flow of water into the drain hole, so, too, would an astronaut be swept by the flow of space into the singularity.

The crucial point is that around the singularity at the center of a black hole is a spherical surface that marks a point of no return. Anything that crosses this surface will never come back out. Nothing inside it, not even a beam of light, can emerge from within it. If an astronaut falling toward the singularity shines a beam of light back toward the spherical surface, that light beam will fall back toward the singularity because space is flowing inward faster than the light can travel outward through that space. This spherical surface is a one-way crossing from the space outside into the space within it. Physicists call this surface the *event horizon* of the black hole.

Before exploring the story of how the event horizon links to thermodynamics, I should stress that the existence of black holes has in recent years been confirmed. Proof comes from the fact that even outside their event horizons, space and time are warped in such a way that the movements of nearby stars are affected. That's why astronomers have observed stars orbiting invisible objects in many different places in the cosmos. The most plausible explanation for such behavior is that the stars are orbiting black holes.

Evidence for black holes also comes from the fact that according to the general theory of relativity, when two black holes collide, they coalesce into a single hole and release a great deal of energy in the form of ripples or waves in space. Remember, space behaves like a liquid, so waves can form in it in a way that's analogous to the way waves form in water. If two icebergs collide in the ocean, that causes ripples to spread outward from the point of the collision through the water. Similarly, when black holes collide, that causes ripples to spread outward through space. These waves were found and measured by two specially built detectors in North America in 2015.

Let us return to a black hole's event horizon, this strange one-way door out of our region of space, because scientists encountered here the strangest incarnation of the laws of thermodynamics. The story of how this was discovered begins with one of the few people in history who exceeds the hype that surrounds him, Stephen Hawking.

In the summer of 1962, a healthy young student sat before a board of examiners at Oxford University. Twenty years old, Stephen Hawking was about to receive his undergraduate degree. As his physics tutor at the time said of the examiners, "They were intelligent enough to realize they were talking to someone far cleverer than most of themselves." Shortly after, Hawking moved from Oxford to Cambridge to pursue his PhD, and he would remain there for the rest of his life. Not long after the move, however, Hawking began showing symptoms of motor neuron disease, the cruel degenerative disease that condemned him to a life of ever-increasing immobility. It began with his losing the ability to walk and thereafter to not being able to feed himself. Eventually he had to breathe through a tube inserted in his throat. The story of how Hawking overcame the initial shock of the diagnosis and the immense difficulties caused by this disability to carry out groundbreaking work as a theoretical physicist is rightly celebrated as inspiring.

During the late 1960s and early 1970s, Hawking collaborated with the brilliant Oxford-based mathematical physicist Roger Penrose. They jointly advanced the understanding of how general relativity shaped the early formation of the universe and investigated many aspects of black holes. By 1970, this work had led Hawking to an uneasy awareness that black holes might have a relationship with thermodynamics. This stemmed from work he'd carried out to show that there was no obvious way for the event horizon of a black hole to become smaller.

For the following reason: All objects, from stars and planets to passing spaceships, can fall into a black hole, adding to its mass. As this happens, its pull on the flow of the space around it increases. Therefore, the speed of the "space flow" reaches light speed at a greater distance from the center of a black hole as its mass goes up. The radius of the event horizon grows. But nothing can fall out of a black hole and reduce its mass. Therefore the radius of its event horizon cannot shrink. Hawking spotted an uncanny similarity between this behavior and the behavior of entropy. Both event horizons and entropy never decrease.

But this similarity, Hawking felt, was a coincidence. It was impossible, he believed, for the event horizon to have anything to do with entropy for the simple reason that all objects that have entropy are warm. Think of a box of gas. If it has entropy, that means that its constituent atoms are moving between various indistinguishable states. That was how Ludwig Boltzmann and Josiah Willard Gibbs had defined entropy. The only way for the gas to have zero entropy would be if its molecules were fixed and motionless. But by definition that would also mean their temperature was absolute zero. The point is, if molecules have entropy, they are moving, and they therefore have a temperature. By this reasoning, for a black hole to have entropy, it must, like a gas, have a temperature. And that in turn means that it must radiate heat. But this appears impossible because nothing, including heat, can escape the event horizon.

In 1971, Hawking published a paper that admitted there was a similarity between the area of the event horizon and entropy—they appeared to be the only things in the universe that inevitably increase and cannot become smaller. But this was merely a coincidence, Hawking argued, because a black hole cannot radiate heat and cannot therefore have entropy.

Unbeknownst to Hawking, however, the previous year a conversation had taken place in an office at the Institute for Advanced Study in Princeton, New Jersey, which hinted that might not be true. The participants were a young PhD student named Jacob Bekenstein and his supervisor, John Wheeler.

John Wheeler was a contradiction. On the one hand, he was a conservative, anti-communist patriot who worked to develop America's nuclear arsenal; on the other, his friends included Soviet scientists and Chilean

Communists. Nearly always dressed in a suit and exuding the air of a corporate executive, he approved of the civil rights movements, of women's rights, and the increased tolerance of diversity that symbolized the 1960s. One of his thumbs was missing a chunk after a youthful experiment with fireworks had gone wrong. As an adult, he would signal his boredom at science conferences by bursting puffed-up paper bags. As well as being one of the twentieth century's most insightful thinkers, Wheeler was also the person who popularized the term *black hole*. Previously black holes had been called gravitationally collapsed objects or Schwarzschild singularities. In many ways, Wheeler was the twentieth-century equivalent of the great Berlin-based physics teacher Gustav Magnus, in whose house in the 1840s the young Rudolf Clausius had begun his investigations into entropy. In Wheeler's study in Princeton over a century later, in late 1970, the problematic nature of entropy was still on the agenda as he sat talking to a young graduate student, then only twenty-three years old, named Jacob Bekenstein.

Bekenstein had traveled an unlikely path to Wheeler's office at Princeton. Born in 1947 in Mexico City, he was the son of Jewish refugees who had fled Europe in the 1930s. Bekenstein's father was a carpenter and his mother a homemaker. Having had to remake their lives, they were not wealthy, yet they now nurtured their sons' talents. As a child Jacob had been taken by his mother to the main library in Mexico City, so he could read books not available at school and develop his growing interest in science and technology. Fascinated by the Soviet *Sputnik* rocket programs of the 1960s, he and his school friends built homemade rockets with fuel they concocted from chemicals bought with pocket money from a medical-supplies store. "A few actually flew," wrote Bekenstein later, "and in a couple of cases so far that we never recovered them." Then in the early 1960s, the Bekensteins secured permission to move to the United States and caught a bus from Mexico City to Texas. The family eventually settled in New York City, where Bekenstein finished high school and did so well as an undergraduate that he won a scholarship to pursue a PhD at the Institute for Advanced Study at Princeton. By this time, Bekenstein had decided on a career as a theoretical physicist, later explaining, "As a young man, while weighing the prospects of various possible professions, I decided I wanted to do something which would be comprehensible even to beings coming from another part of the universe."

On that day in 1970, as Wheeler sat in his Princeton office with Bekenstein, the two men were mulling over the same concern about energy dissipation that had so fascinated Carnot, Kelvin, and Clausius. What those nineteenth-century scientists had highlighted was that when a temperature differential exists in a system, meaning when one part of it is hotter than another, then—and only then—can useful work be extracted from heat. So, if one imagines, as Lord Kelvin did, an iron bar that's hot at one end and cold at the other, the flow of heat from hot to cold can be exploited to perform useful mechanical work such as raising a weight. If, however, the heat is allowed to dissipate through the iron bar, so that the same temperature is established throughout its length, the heat is now useless, even though no energy has been destroyed. It can no longer be used to lift a weight. The iron bar has gone, in other words, from a low- to a high-entropy state. The idea that by increasing entropy the opportunity to extract useful work from energy is lost prompted Wheeler to say to Bekenstein, "I always feel like a criminal when I put a cup of hot tea next to a glass of iced tea and then let the two come to a common temperature, conserving the world's energy but increasing the world's entropy. My crime echoes down to the end of time for there is no way to erase or undo it. But let a black hole swim by and let me drop the hot tea and cold tea into it. Then is not all evidence of my crime erased forever?" To most other listeners, this statement would have been mystifying, but as Wheeler said, "This remark was all that Jacob needed."

That comment inspired Jacob Bekenstein's PhD thesis, which can now be seen as the first step toward an audacious and still unresolved new physics. What's remarkable is that Bekenstein felt that the principles of thermodynamics, born in the everyday technology of steam engines, could not be ignored. "I was very dissatisfied with this conclusion," he later wrote of Wheeler's remark, for "the second law of thermodynamics is so general, works in so many cases, that I was not ready to swallow its becoming irrelevant."

For another way to see what troubled Bekenstein, imagine a box full of hot gas, which, of course, has entropy. Now, let's drop the box past the event horizon of a black hole. Because nothing can come back from across the event horizon, the box has crossed a point of no return and is thus no longer part of our universe. Both the box of gas and the entropy associated with it have disappeared from our universe. But that means that *the entropy of our universe has gone down*, which directly contradicts the second law of thermodynamics. When it came to black holes, there seemed to be a conflict between general relativity and thermodynamics, with the former coming out on top.

Bekenstein decided to see if thermodynamics could somehow survive the battle with general relativity. To do so he had to assume, flying in the face of the beliefs of Stephen Hawking and others, that a black hole could have entropy. Colleagues have commented how Bekenstein's quiet and gentle demeanor stood in stark contrast to his intellectual boldness. "In fact, Bekenstein's style of doing physics was much like Einstein's," wrote the physicist Leonard Susskind, "both were masters of the thought experiment. With very little mathematics, but with a lot of deep thinking about the principles of physics and how they apply to imaginary (but possible) physical circumstances, both men were able to draw far-reaching conclusions that would profoundly affect the future of physics."

Reconciling thermodynamics with black holes certainly needed "deep thinking." To do so, Bekenstein had to combine the theories of relativity, thermodynamics, information theory, and a smattering of quantum mechanics. His logic went as follows:

Bekenstein first asked, What's the smallest amount of entropy that I can add to a black hole?

His answer: the smallest quantity of energy that can be dispersed throughout the space inside the black hole's event horizon. Thomson would have approved; Bekenstein was essentially thinking of the space within the event horizon as if it were an iron bar in which all the heat is evenly spread.

So, what is the smallest amount of energy that can be dispersed or spread out inside the event horizon? Bekenstein imagined this to be a single photon of light that could be anywhere inside the event horizon of a black hole—a single photon, in other words, whose wavelength was roughly equal to the radius of a black hole's event horizon. To calculate how much energy it would have, Bekenstein turned to Einstein's 1905 paper on light quanta, which said that the energy of photons was proportional to their wavelength. This allowed Bekenstein to estimate the size of this minimum dispersed energy, arriving at the answer that it was proportional to the radius of the black hole's event horizon.

Knowing the amount of energy dispersed inside the event horizon, Bekenstein converted it into mass using Einstein's famous formula, $E = mc^2$.

This was a crucial step. It told Bekenstein that increasing the entropy of a black hole increased its mass. Per the general theory of relativity, increasing the mass of a black hole always increased the area of its event horizon. This was also in line with Stephen Hawking's recent paper showing that event horizons can never become smaller.

To summarize: Entropy increases the energy content of the black hole, increasing both its mass and the size of its event horizon. Bekenstein's argument? *Whenever the entropy of a black hole increases, so does the area of its event horizon.* In other words, the area of the event horizon of a black hole was not an analogy for entropy, *it was a direct measure of its entropy.* In Bekenstein's view, this saved the universal applicability of the second law of thermodynamics. The entropy of the universe always increases, even when things fall into black holes, because the entropy lost from the space outside the event horizon is made up for by an increase in the surface area of the event horizon. Bekenstein called this the generalized second law of thermodynamics, or GSL.

Bekenstein presented GSL as his PhD thesis to Wheeler a few months after they'd first discussed the topic. Of his response, Wheeler later said, "I had learned often enough in my career that nature has a way of being a little stranger than we think it ought to be. So I said to Jacob, 'Your idea is so crazy that it might just be right. Go ahead and publish it.'" Bekenstein did so.

Yet when Bekenstein's paper came out in 1972, few took it seriously. Yes, Bekenstein had shown a mathematical link between the entropy of a black hole and the area of its event horizon, but he had ignored that this entropy implied that a black hole had to radiate heat. No one believed that possible. "Those were two lonely years," remembered Bekenstein in his autobiography. "Black hole entropy was very new at the time, and most people who heard about it said it was patent nonsense. Some told me that I was wasting my time."

Reading Bekenstein's paper did not make Stephen Hawking happy, either. Having spent several years studying general relativity, Hawking felt strongly that it banned the possibility of black holes giving off heat. Together with two colleagues, he immediately wrote a new paper pointing out why Bekenstein was wrong. Particularly annoying to Hawking was that his work was cited by the Princeton physicist. "I must admit that in writing this paper I was motivated partly by irritation with Bekenstein, who, I felt, had misused my discovery of the increase of the area of the event horizon," wrote Hawking in his bestseller *A Brief History of Time.*

A year later, events took another surprising turn. In September 1973, Hawking visited Moscow and discussed black holes with two of the Soviet Union's leading physicists, Yakov Zeldovich and Alexander Starobinsky. On returning to England, Hawking felt the ideas raised in these conversations would prove that a black hole cannot radiate heat and therefore cannot have entropy. However, as Hawking began his calculations, he found to his "surprise and annoyance" that his results seemed to be suggesting the opposite of what he'd hoped. As he wrote, "I was afraid that if Bekenstein found out about it, he would use it as a further argument to support his ideas about the entropy of black holes, which I still did not like." But the more Hawking worked, the more he seemed to be proving Bekenstein right. Not only did black holes radiate heat, but they did so by exactly the amount required if the area of their event horizons was indeed a measure of their entropy. By early 1974, Hawking had developed this work into a fully fledged theory. It led to his now-famous discovery that "Hawking radiation" leaks out of all black holes.

But how did Hawking realize that although nothing—not even light— can escape the event horizon, a black hole can radiate heat in defiance of this principle? The answer lies in Hawking's having decided to investigate the event horizon of a black hole from the perspective of quantum theory. At the time, most physicists felt that black holes, massive cosmological objects that obeyed the principles of general relativity, had little to do with quantum theory. After all, quantum theory is a guide to the microscopic world inside the atom. However, Hawking had a hunch, in part inspired by his conversations in Moscow, that if he investigated the empty space at and near the black hole's event horizon from the quantum perspective, something interesting would turn up. To follow Hawking's exact reasoning is tricky. For a rough intuition for what he did, we must consider one of the most outlandish consequences of the famous "uncertainty principle" of quantum physics—something known as "vacuum energy."

As the name implies, far from being inert, the vacuum is seething with activity. At any instant, bursts of energy appear from nowhere by borrowing equivalent bursts of energy from a tiny instant in the future. Mostly, we are unaware of these fluctuations because the positive burst of energy that appears at one instant is cancelled out by the negative burst that immediately follows it. Negative energy is a strange concept, but it does exist! These energy bursts can appear in many forms. They can appear as particles, such as electrons and positrons, and as photons of electromagnetic energy.

Hawking surmised that on and just outside the event horizon of black hole this "cancelling out" is disrupted. The extreme curvature of space and time here means some of the negative energy created is torn away from the positive energy that it normally would have annihilated. This positive energy, having survived, is free to radiate away from the black hole. The negative energy falls into it. And because it's *negative* energy, it has the effect of making the black hole *less* massive.

To an outside observer, it appears as if the black hole is "evaporating," slowly shrinking as it emits energy—this is what is referred to as "Hawking radiation."

What's utterly remarkable about Hawking's calculations was they enabled him to predict the temperature of this radiation emanating from the event horizon. It's typically very low—a tiny fraction of a degree above absolute zero. But it is exactly what one would expect if, as Bekenstein had claimed, the entropy of a black hole is proportional to the area of its event horizon. "It turned out in the end that he [Bekenstein] was basically correct, though in a manner he had certainly not expected," wrote Hawking later.

A rabbit had been pulled out of the physics hat. Between them, Hawking and Bekenstein had shown that the three great ideas of modern physics—general relativity, quantum mechanics, and thermodynamics—work in harmony. For these reasons, black hole entropy and radiation have come to dominate contemporary physics as scientists search for a so-called grand unified theory, the Holy Grail of a single principle that explains nature—the world, the universe, everything—at its most fundamental level.

In the decades since Hawking's and Bekenstein's work, a consensus has formed that the surface area of a black hole's event horizon *is* its entropy. This odd notion hints at something fundamental about how our universe is structured. Entropy is normally considered a three-dimensional phenomenon. The entropy of a box of hot gas, for instance, is the sum of all the different but indistinguishable ways that the atoms of the gas within the box can arrange themselves in three spatial dimensions. So how can this patently three-dimensional process turn into a two-dimensional one on the surface of the event horizon?

Sparking intense research over the last few decades, another field of physics entered the discussion along with thermodynamics, quantum theory, and general relativity. That field is information theory. To see why, imagine again that a box of hot gas is hurtling into the black hole. To calculate the gas's entropy, one could theoretically make an enormously long list of the position of each gas molecule and the direction in which it was traveling. One could then convert each item in this list into a binary number using the methods discovered by Claude Shannon in the 1940s. Doing so would generate a complete description of the gas's contents as a long string of 1s and 0s. With this in mind, remember that according to Bekenstein and Hawking, when the box of gas falls into the black hole, the area of its event horizon increases by an amount determined by the gas's entropy. It's as if the surface area of the black hole's event horizon increases by an amount just big enough for all the 1s and 0s that describe the gas's entropy to be encoded on it.

Using Bekenstein's and Hawking's formulas, physicists can say how much space a single digit from the binary number describing the gas's entropy takes up on the event horizon's surface. It's a small area—around $4 \times 10^{-66}$ square centimeters. This means you can think of the surface of a black hole's event horizon as covered in tiny tiles, each carrying one "bit" of the information describing the entropy of everything that's fallen inside.

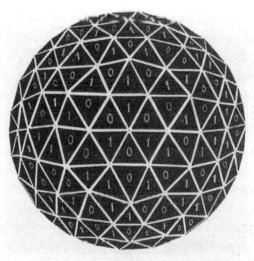

Each triangular tile contains one bit of information. Collectively, they describe the entropy of everything inside the black hole.

The mathematics shows that the event horizon's area grows the right number of tiles to encode every "bit" of entropy that falls into the black hole. For a rough visual analogy, imagine a stream of oil pouring onto a sphere, coating it in a very thin layer. As more oil flows, the sphere expands by an amount which ensures that the oil layer remains infinitesimally thin. Similarly, to observers outside a black hole, objects don't appear to fall into it. Rather, they are smeared in a thin layer onto the event horizon.

This has led physicists to compare the event horizon of a black hole to a hologram. These are two-dimensional surfaces that contain all the information needed to generate a complete three-dimensional image. These aren't like the 3D images one sees in the cinema, which create an illusion of three dimensions. One can walk in a circle around a hologram of an object and it will appear as if one is walking in a circle around the actual three-dimensional object. Yet all the information needed to create the image is stored on a flat piece of film. This so-called holographic principle has led physicists to suggest that the two-dimensional information on the black hole's event horizon is in a sense more "real" than the three-dimensional stuff that has fallen into it because the event horizon is still accessible to our part of the universe while whatever has fallen in is lost forever.

And the conclusion this brings us to is even more extraordinary: it could be that all the information that describes our universe is stored on the two-dimensional surface shell that surrounds it.

The reason is that in 1998 it was discovered that the rate at which the universe is expanding is accelerating. The speed at which distant galaxies recede from us is growing, becoming ever faster. Physicists do not yet know what is behind this expansion—a mysterious "dark energy" has been suggested—but the effect is that space is flowing away from us, outward in all directions, at greater and greater speeds, the farther away we look. At a distance of around 16 billion light-years from earth, space is flowing away from us at a speed faster than that of light. That means that as galaxies cross this boundary, they become lost to us; their light can never travel back fast enough to reach us. Assuming that the universe's acceleration continues indefinitely, more and more galaxies will be swept away from us by space flowing faster than the speed of light, and they will disappear from view. Eventually, our own galaxy and a few of its closest neighbors will be all that's visible.

One way to visualize this is to imagine that all the galaxies in the universe are dots on the surface of a spherical balloon. A mysterious energy is making the balloon expand at a faster and faster rate, so all the dots are moving farther and farther apart.

Think of one dot as our Milky Way, and imagine that a circular boundary is drawn around it and its nearby galaxies. Inside this boundary, the balloon is expanding slower than the speed of light, whereas outside it is expanding faster. Only objects inside this boundary are visible to us; everything outside the boundary is invisible. But because the rate at which the balloon is expanding is increasing, more and more of these objects will disappear as they flow across the boundary.

This process should sound familiar. It's as if our universe is "an inside-out black hole." Instead of stuff flowing *inward* across a one-way boundary never to be seen again, stuff is flowing *outward* across a one-way boundary never to be seen again. That means that, like the event horizon surrounding a black hole, there is an event horizon around our universe. The denouement of this argument? Just as all the information needed to describe everything that's fallen into a black hole is encoded on the surface of its event horizon, it could be that all the information needed to describe everything that exists in our universe is encoded on the two-dimensional surface of the horizon that surrounds it. That would suggest that the three-dimensional universe that we perceive is an illusion. It's just the way we experience the real two-dimensional universe. The universe we see is like a hologram, a three-dimensional shadow of a two-dimensional reality.

And the reason that this idea attracts physicists? It hints that not only is the third dimension an illusion, but so, too, is gravity. This force, which all of us experience as real, could be an artifact, a way to help us interpret the two-dimensional data stored on the boundary of the universe. And if gravity is not "real," that means that it does not have to be synthesized with the other forces of nature. It could therefore be that the "theory of everything" that physicists so desperately crave is already in our grasp!

The story of thermodynamics began two hundred years ago with a young Frenchman, Sadi Carnot, who simply sought to make steam engines more efficient. Dying of cholera in a mental asylum, he had no idea that his work would have any consequence. Never, not even in his wildest and most fevered imaginings, could he have foreseen that the ideas he seeded would one day help us to understand the very edge of our cosmos.

As Stephen Hawking wrote, "We are just an advanced breed of monkeys on a minor planet of a very average star. But we can understand the Universe. That makes us something very special."

# Epilogue

At its heart, this book celebrates thermodynamics and its importance to fundamental science. But understanding energy, entropy, and temperature and the laws they follow has played a vital role in promoting the greatest improvement in the human condition in our species' entire history.

Before 1850, most people had short, brutish, disease-ridden lives. People struggled to survive with little more than muscle power—their own and that of domesticated animals. Elites lived well, but they did so by living off other people's muscles.

Then something changed. Thanks to the scientific breakthroughs described in this book, muscle power was, over time, replaced by other sources of power, such as coal, oil, gas, hydro, and nuclear. The result is that, by and large, we now live longer, happier, healthier, and more fulfilling lives than at any other point in our history. This "good news" story is often ignored, but if you need data to persuade you, I'd recommend a website created by the Oxford-based economist Max Roser called ourworldindata. Exhaustive evidence there from all over the world makes this point.

In short, I believe that the discovery of the laws of thermodynamics is one of the most consequential and beneficial advances ever made. But, rightly, some readers might argue that in my enthusiasm for scientific and technological progress, I've ignored the damage that industrialization causes to the environment. It's a fair point.

First, no sensible environmentalist, however concerned about the state of the planet, wants humanity to return to the world of the early 1800s. That's an event horizon too far, for no one wants a return of the poverty, the disease, and the appalling rates of child mortality that dominated human existence for most of its history. Second, most would agree that it's science and technology that's driven this improvement in the human condition. But the

241

question is, Will climate change, a product of that same science and technology, overwhelm us, undoing all the progress we've made as a species?

Here, I'd like to tell a story. It's about another great Victorian scientist—John Tyndall.

A gifted experimental scientist with a passion for engaging the general public, Tyndall was born in 1820 in county Carlow in Ireland. The family were Anglo-Irish, and Tyndall's father was the local police constable. In his twenties Tyndall left home to work as a surveyor on the rapidly expanding railway network in England. Despite little formal scientific education, he then developed a burning interest in physics. The best way to pursue that, Tyndall decided, was to study in Germany. The English universities' reputed obsession with the classics and pure mathematics, he felt, meant that their standards of laboratory and experimental science were poor. By contrast, these were the very aspects on which Germany's new universities placed a high premium.

So from 1848 to 1851, Tyndall lived, worked, and studied in the town of Marburg, where he forged close relationships with some of Germany's leading experimentalists. Not only did he study under the great chemist Robert Bunsen, of "burner" fame, but Tyndall was also the first to translate Rudolf Clausius's writings on thermodynamics into English. Additionally, Tyndall fell in love with the Alps. He pioneered mountaineering in this range and was among the first climbers to reach the peak of the Matterhorn.

By 1851, when Tyndall returned to Britain, he had become one of the most accomplished experimental physicists in the country. These skills brought him to the attention of the head of magnetic investigations at the Royal Institution in London, Michael Faraday, through whom Tyndall became the institution's professor of natural philosophy. This meant that Tyndall now had access to the best-equipped physics laboratory in the British Isles, a resource he used with great imagination and determination.

Tyndall studied a great many scientific topics ranging from magnetism to sound to how to render milk safe to drink by heating it. But his greatest work concerned the earth's atmosphere and its ability to absorb, radiate, and retain the heat that pours into it from the sun. In essence, Tyndall wanted to investigate the role that the atmosphere plays in maintaining the earth's temperature.

To that end, Tyndall devised a beautiful and historic experiment, which he carried out in the early 1860s.

Tyndall knew that the sun's energy reaches us mainly as visible light, which in turn warms the soil and the water on the earth's surface. These then radiate some of that energy back into the atmosphere. But the form of the energy that's radiated back out of the earth is rather different than visible light. This emerges as infrared radiation. At a lower frequency than visible light, it's invisible but feels warm.

What Tyndall wanted to know, however, was how infrared radiation behaves in the earth's atmosphere. Does it, like its visible counterparts, travel unhindered through the air? Or is it, to some extent, trapped or blocked by it? This was an important question, because if it couldn't escape back into space, this infrared radiation would act to warm the atmosphere.

To find out if this was the case, Tyndall first needed a source of infrared heat in his laboratory. After much trial and error, he settled on a copper cube filled with boiling water. He positioned it at one end of a horizontal four-foot-long tube that could be filled with gas. At the other end he placed a "thermo-electric pile," which acted as a sensitive thermometer. This told Tyndall if any of the heat radiating out of the copper cube was being absorbed in the gas-filled tube.

Tyndall discovered something quite extraordinary. Infrared radiation was barely affected by nitrogen and oxygen, the gases that make up 99 percent of the atmosphere. But when the same radiation passed through air containing water vapor or carbon dioxide, even though these gases are present in tiny amounts, the thermometer registered a fall in the temperature reading. The only explanation, argued Tyndall, was that the presence of these gases increases the air's ability to absorb infrared radiation by a factor of about fifteen.

Tyndall had discovered what we now call the greenhouse effect. Water vapor and carbon dioxide in our atmosphere trap some of the sun's energy. Effectively, they provide a warming blanket around our planet. Without these gases in our atmosphere, the earth's temperature would plummet and would be well below zero even at the equator.

But what happens if the levels of these greenhouse gases—the water vapor and the carbon dioxide—go up? Clearly, more heat radiating out from the earth will be trapped, and the earth's temperature will rise. As Tyndall and his contemporaries immediately noted, the coal-burning factories of the Industrial Revolution were belching carbon dioxide straight up into the atmosphere, causing the levels of this greenhouse gas to rise.

Thus Tyndall made it clear as long ago as the 1860s that human industrial activity could affect the climate. That is why, as early as 1917, Alexander Graham Bell, the great technologist and inventor of the telephone, was advocating the use of solar power to mitigate the possible dangers of unabated burning of fossil fuels.

I suppose that's the underlying point of this story. As we investigated heat, we learned to harness it, which hugely improved the human condition. But from the very early stages of this story, we have had warnings of the potential dangers—and crucially, we have had time to think about how to alleviate them. Today, thanks in no small part to scientists' understanding of the laws of thermodynamics, we know of several strategies to deal with climate change. Wind farms and other renewable sources already provide around a third of the United Kingdom's electricity, and we have the know-how to ratchet that contribution up. Scientists with impeccable green credentials ranging from James Lovelock to Mark Lynas now argue for much more nuclear power generation on the grounds that it's carbon neutral and considerably safer than most people think. Geothermal and tidal power are also promising. The main obstacle to dealing with climate change isn't scientific. Instead, it's political and emotional. While some refuse to accept that the problem exists, others refuse to accept the solutions.

That brings me back to why I wanted to write this book. Now, more than ever, it's important that all of us have a basic grasp of thermodynamics, so that we can make sensible and informed decisions about how best to ensure progress while preserving or improving living conditions for our fellow humans without ruining the environment. Should we commit to nuclear energy? Should we drive electric cars? How much tax should we pay on petrol, and how much should we subsidize wind farms? We will be in no position to answer these vitally important questions unless we have a basic understanding of the laws of thermodynamics.

Informed debate, I am convinced, will provide these answers. The science of heat can and should improve the human condition without destroying the planet.

It's up to us.

# Acknowledgments

Many people helped me in immeasurable ways as I worked on this book.

I was very fortunate in having an agent, Patrick Walsh, who supported, encouraged, and advised me with great skill from the proposal stage through to final manuscript. A big thanks is due to John Ash at Pew Literary, too. I am very grateful to Myles Archibald at HarperCollins and Daniel Loedel at Scribner, who shared my enthusiasm for this story from the outset. At Scribner, I am also very grateful for the hard work and support of Sarah Goldberg, copyeditor Steve Boldt, and designer Erich Hobbing.

My thanks go to Khokan Giri for creating wonderful illustrations that capture scientific ideas while being both aesthetically pleasing and playful

I have called on the expertise of numerous people as I navigated the historical and scientific complexities of this tale. Dr. Daniel Mitchell, research fellow and director at the Center for Historical Research in Philadelphia, gave me invaluable guidance, especially for the first two-thirds of the book. As I strove to find intuitive explanations for many scientific concepts, I received generous and enthusiastic support from Graham Woan, professor of astrophysics at the University of Glasgow. To help me explain James Clerk Maxwell's experiments on kinetic theory, I am very grateful to Dr. Jim Shaikh. For the chapter on Alan Turing and morphogenesis, I received invaluable advice from Jeremy Green, professor of developmental biology at King's College London. For the chapters on information theory and Maxwell's demon, I am very grateful for comments by Professor Danielle George at Manchester University and by Professor Raja Sengupta at UC Berkeley. For the final chapter on black holes, I turned to Dr. Sownak Bose from the Harvard-Smithsonian Center for Astrophysics. For this chapter, I am also grateful for the time Yehonadav Bekenstein took to talk to me about his father, Jacob.

I am enormously grateful to Simon Schaffer, professor of the history of science at the University of Cambridge. Not only did he patiently comment on and correct many of the pages herein, but also it was his passion for and knowledge of the history of science that in many ways inspired this book.

Above all my gratitude goes to my friend Andrew Smith—for his support, both intellectual and emotional through all the ups and downs of writing a book. Few authors have the privilege of counting as a friend someone with Andrew's impeccable taste, astute judgment, and ability to encourage.

# The Carnot Cycle

Sadi Carnot didn't just want to show that the maximum amount of motive power that can be extracted from a given flow of heat is set by the temperature difference between furnace and sink. He also wanted to deduce the maximum amount of motive power, M, that an engine can produce from a given flow of heat, H. (The ratio M divided by H is the engine's efficiency.) To answer this, Carnot remained in the abstract and described how an ideal engine would work. Such a machine would be unbuildable in practice, but it would define the upper limit of any heat engine's efficiency.

Carnot had already shown that an ideal engine's efficiency is independent of its working material. So, for his argument he proposed a substance that was well understood and that physicists had studied since the seventeenth century—atmospheric air. Its behavior as it is heated, cooled, squeezed, or expanded follows known mathematical laws. Air is like most gases, but because it remains a gas well below 0°C, the laws predicting its behavior as a gas would hold over a wide temperature range. (The same laws break down for steam when it liquefies into water at 100°C.)

Two specific aspects of air's behavior were of particular interest to Carnot. First, the unexpected fact that you can raise or lower the temperature of air without adding or removing heat from it.

Imagine a cylinder like the one in a steam engine, which is positioned vertically so the piston can slide up and down. The cylinder is insulated so no heat can flow in or out. Now press down on the piston, squeezing the air so it occupies, say, half its original volume. The gas will resist—you will have to exert yourself, putting motive power into the cylinder—but as you push down, the air's temperature will rise by around 60°C.

This kind of compression, because no heat flows into or out of the gas, is termed adiabatic. Adiabatic compression can be reversed. Having squeezed the air, let go and allow it to expand and push back the piston to its original position. As long as the cylinder remains insulated, the gas will cool down to the temperature it started at and it will give back the same amount of motive power you put in to squeeze it. Because no heat flowed in or out of the gas, this is known as adiabatic expansion.

The second aspect of air's behavior Carnot focused on was what happens when heat from a furnace *does* flow into a gas. This can both raise its temperature *and* cause it to expand, pressing on the walls of its container. This expansion is what pushes the piston of a steam engine. Carnot concluded that for maximum efficiency, *all* the heat flowing into the cylinder should go to expanding the gas and *none* to raising its temperature.

But this seems contradictory—how can you add heat to something without making it hotter?

It's almost impossible to achieve in practice, but theoretically, this is how: A piston is positioned almost at the very bottom of a vertical cylinder. Hot air at the same temperature as an adjacent furnace is squeezed into the small space between piston and cylinder floor. The gas expands and pushes the piston, thus creating motive power. If the cylinder was completely insulated—that is, adiabatic—the gas would cool. But as the cylinder is next to the furnace, heat will flow into it, compensating for any drop in temperature.

So, as heat flows into the air, it expands, creating a quantity of motive power while its temperature remains unchanged. This is known as an isothermal expansion. For a given, fixed temperature, it creates the most motive power possible from a given amount of heat.

Like the adiabatic process, the isothermal one can run backward, too. In this case, you press down on a piston in a cylinder containing gas at the same temperature as an adjacent sink. The temperature does not rise as it would in the adiabatic case because as it starts to do so, the heat flows into the sink. This is known as isothermal compression. At a given, fixed temperature, this uses up the *least* amount of effort or motive power to compress a gas, while removing heat from it.

With these two processes, adiabatic and isothermal, in mind, Carnot sketched the ideal, maximally efficient heat engine.

The picture shows a single vertical cylinder containing a piston that moves up and down. Below it on the left is a furnace, labeled A, on the right a sink, labeled B. Heat is used to expand gas, and this pushes the piston directly.

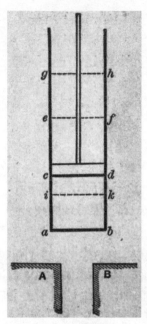

Carnot's original diagram
of an ideal engine

When the gas needs to be heated, the cylinder containing it is brought into contact with the furnace (A), and when the gas needs to be cooled, the cylinder moves across and is brought into contact with the sink (B). Carnot imagined the furnace to be so vast that no matter how much heat flows out of it, its temperature doesn't fall. It remains at, say, T(furnace) degrees. Similarly, he imagined the sink to be so large that no matter how much heat flows into it, its temperature does not rise, remaining at, say, T(sink).

Then Carnot describes how the engine functions. It works in a four-step loop that repeats over and over.

## Step 1

The piston is positioned almost at the very bottom of the cylinder, and a quantity of hot air at the same temperature T(furnace) as the furnace is squeezed into a small volume between the piston and the cylinder's floor. The furnace is brought into contact with the cylinder, and a quantity of

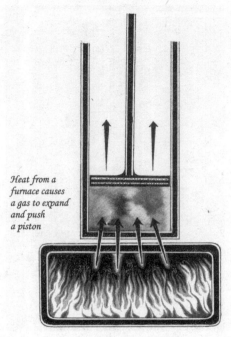

*Heat from a
furnace causes
a gas to expand
and push
a piston*

Step 1 of Carnot's cycle

heat, call it H, flows into the gas. This expands and pushes the piston, producing an amount M(1) of motive power.

Carnot stipulates this is an isothermal process, so all the heat H goes to producing motive power M(1).

This is called the power stroke, as most of the motive power of the engine is realized at this stage.

But for an engine to be useful, it can't stop here. The piston needs to go back down to the bottom of the cylinder, so the process can repeat.

### Step 2

For the cylinder to go back down, the gas that was pushing the piston must be squeezed back down. The best way to do that, Carnot argued, is to make the gas as cold as possible. That's because it takes less effort to compress cold gas than hot gas.

(If you're not convinced of this, blow up a balloon and put it in the fridge. Within a few minutes, it will have shrunk down to a fraction of its former size as the air inside it cools and becomes more "squeezable." It's also why car tires sag in the winter.)

But what's the best way of quickly cooling a gas in the cylinder of an engine? The answer is to allow it to expand adiabatically, pushing the piston even farther up.

In this step, therefore, the engine creates a little more motive power, call it M(2), and the gas in the cylinder cools down to a temperature of T(sink).

## Step 3

With the gas now at a much-lower temperature and thus much more squeezable, some fraction of the motive power M(1) created in Step 1, call it M(3), is used to push the piston down and compress the gas back to a small volume.

Carnot stipulates that this compression is isothermal, so M(3) is as small as possible.

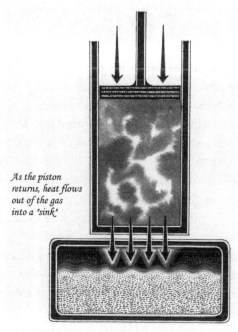

*As the piston returns, heat flows out of the gas into a "sink"*

Step 3 of Carnot's cycle

(As stipulated by caloric theory, Carnot believed that during this compression, all the heat H that entered the gas in Step 1 flows out of it into the sink.)

By the end of this step, the gas is squeezed almost completely back into the tiny space it started out in. But an amount M(3) of motive power has been used up.

### Step 4

Now the gas must once again be heated to the temperature of the furnace in readiness for Step 1 to be repeated.

But if heat from the furnace is used for this, it would be wasted, as it isn't creating motive power. So, the cylinder is once again insulated, and some more motive power, M(4), is used up pushing the piston back down, squeezing the gas. Because no heat can leave the cylinder, this adiabatic compression forces the temperature of the gas back up to T(furnace).

The gas is now back in the same state it was at the start of Step 1—hot and ready to expand and push the piston all over again.

This step is the precise opposite of Step 2. The motive power used up in Step 4, M(4), is the same as M(2), the amount created in Step 2. So, they cancel out.

This four-step loop, known to engineers and physicists around the world today as the Carnot cycle, is one of the great thought experiments in science. It allowed Carnot to assess the theoretical maximum amount of motive power obtainable from a supply of heat.

During the cycle, an amount H of heat flowed into the gas from the furnace and left it for the sink. The net amount of motive power produced? M(1) minus M(3), the difference between the amount created in Step 1 and used up in Step 3. So, the efficiency of an ideal engine? M(1) minus M(3) divided by H.

The route to improving engine efficiency was clear. Make the gas as hot as possible as it expands and as cold as possible when it's compressed. Gases expand more forcefully the hotter they are and are easier to compress the colder they are. Therefore, the bigger the difference in temperature at which these steps occur, the more efficient the engine.

# How Clausius Reconciled the Conservation of Energy with the Ideas of Sadi Carnot

Rudolf Clausius believed that heat and work do turn into each other *and* heat must flow from hot to cold for work to be created. He provided the first correct analysis of the relationship between heat and work. His ideas live in every gas, diesel, and jet engine, as well as the steam turbine and rocket.

Clausius started by revisiting Carnot's ideal engine and its four-step cycle, but with the idea of work energy interconvertibility in his armory. In so doing, Clausius identified a form of energy that's hidden from view, which he represented by the letter $U$. Today, this quantity is usually referred to as internal energy.

Imagine a balloon that's full. The high-pressure air trapped inside, pushing against the skin of the balloon, is a store of energy, rather like a battery is a store of electrical energy. Just as the battery's charge can be used or replenished, so can the internal energy of the gas.

Squeeze the balloon with your hands and it will resist and become hotter. This shows that the effort you're putting in as you squeeze—the work you're doing—is further increasing the internal energy of the air inside.

Now cup the balloon in your hands so it can't expand and heat it. You'll feel the expanding pressure and temperature of the balloon rise—signs that the heat you're adding is turning into the gas's internal energy, boosting its level.

Internal energy can be released as heat. A filled balloon placed in a cool environment such as a fridge will, as it gives up its internal energy as heat to its surroundings, shrink and become colder.

Internal energy can also be turned back into work. Burst the balloon. Some of its internal energy manifests as the sound of the bang, and some manifests as fragments of rubber flying across the room and the surrounding air being pushed aside.

In a heat engine, as Clausius saw it, the internal energy of a gas should create work as efficiently as possible. This is how, when applied to Carnot's four-step cycle.

Step 1: Isothermal Expansion

A quantity of hot gas is squashed in a small space between a piston and the floor of a cylinder. It expands, pushing the piston and doing work, hence giving up some of its internal energy. But because it's placed next to a furnace, heat flows into the gas, replenishing that internal energy. This maintains the gas's temperature at a constant value.

So, in this isothermal step a quantity of heat $H(1)$ is turned into an amount of work $W(1)$.

Step 2: Adiabatic Expansion

The cylinder is insulated. The gas inside continues to push the piston,

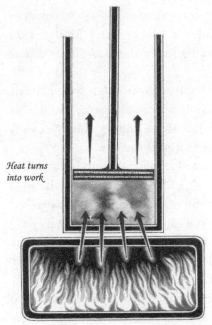

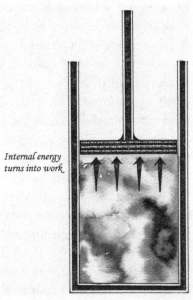

*Heat turns into work*

*Internal energy turns into work*

Step 1: Isothermal Expansion                    Step 2: Adiabatic Expansion

doing work. The gas loses internal energy, but because of the insulation, no heat flows replace it. At the end of this adiabatic-expansion step, the gas is cooler and it has done W(2) of work.

Step 3: Isothermal Compression

The cylinder is placed next to a sink and the gas is squeezed. Work is done *to* the gas. If the cylinder were still insulated, this would increase its internal energy, making it hotter. But because it's next to the sink, this absorbs all the heat created. So, the temperature of the gas does not change. An amount of work W(3) is turned into heat H(3).

Step 4: Adiabatic Compression

The cylinder is insulated, and the piston is pushed down until the gas occupies the same volume it did at the beginning of Step 1. More work is done to the gas, raising its internal energy so its temperature is also the same as it was at the beginning of Step 1. The work done to the gas in this step, W(4), is exactly canceled by W(2), the work done by the gas in Step 2.

For an overall picture, Clausius then totted up the heat flowing in and out of the engine and the work done by it and to it. The table on page 256

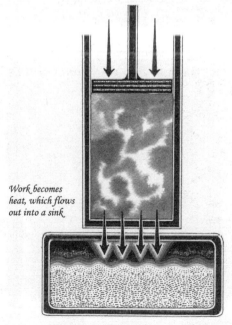

*Work becomes heat, which flows out into a sink*

Step 3: Isothermal Compression

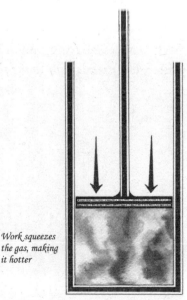

*Work squeezes the gas, making it hotter*

Step 4: Adiabatic Compression

shows this. (Following convention, I've marked heat flowing *into* an engine and work *done by* an engine as positive. The opposites as negative.)

| Step | Heat Flow | Work |
|---|---|---|
| 1: Isothermal expansion | H(1) | W(1) |
| 2: Adiabatic expansion | 0 | W(2) |
| 3: Isothermal compression | –H(3) | –W(3) |
| 4: Adiabatic compression | 0 | –W(4). This is the same size as W(2). |
| Heat converted into work | H(1) – H(3) | |
| Heat lost to sink | H(3) | |
| Net work produced by engine | | W(1) – W(3) |

By analyzing the heat flow through an ideal engine and the work it produces, Clausius reconciled Carnot's and Joule's seemingly contradictory ideas. He came to the following conclusion as to what happens in an ideal engine:

Some heat is converted into work as Joule believed.

The rest of the heat flows into the sink as Carnot held.

The key lesson? You cannot turn all the heat into work; *some will be wasted, lost irretrievably to the sink.*

# The Four Laws of Thermodynamics

This book focuses on the discovery and consequences of the first and second laws of thermodynamics. Over the twentieth century, scientists added two more laws for the sake of completeness. The first of these, now known as the zeroth law, was assumed to be true in the nineteenth century but didn't then have the status of a law. The second of the additional laws, now known as the third law, pertains to materials at ultracold temperatures very near to absolute zero.

### The Zeroth Law
If two thermodynamic systems are each in thermal equilibrium with a third one, then they are in thermal equilibrium with each other.

(Think of this law in terms of thermometers. If such a device obtains the same reading from two separate objects, then heat will not flow between them even if they are brought into contact.)

### The First Law
The energy of the universe is constant.

### The Second Law
The entropy of the universe tends to increase.

### The Third Law
The entropy of a system approaches a constant value as its temperature approaches absolute zero.

(This law allows entropy to be expressed on an absolute scale, rather than only as a change in its value.)

# Notes

## Prologue

ix  *the Cinderella of the sciences*: "Bluff Your Way in the Second Law of Thermodynamics" by Jos Uffink, Department of History and Foundations of Science, Utrecht University (2000).

x  *took an approach to physics inspired by thermodynamics*: See quotes from Einstein on pages xxi and xxii of the introduction to *The Collected Papers of Albert Einstein*, vol. 2.

x  *"It is the only physical theory of universal content"*: See above.

xi  *while producing a television documentary*: Order and Disorder, episodes 1 and 2, first broadcast by the BBC in October 2012.

xi  *"that little essay was"*: From "On the Dissipation of Energy" by William Thomson in *Fortnightly Review*, March 1892.

xi  *"It must be splendid to"*: From "A German Professor's Journey into Eldorado" by Ludwig Boltzmann (*Populäre*, 1905).

## Chapter One: A Tour of Britain

1  *Jean-Baptiste Say*: See "De l'Angleterre et des Anglais: l'expertise de Jean-Baptiste Say de l'industrie anglaise," *Innovations* 45 (3) (2014), by André Tiran.

2  *Britain's population had soared*: Data on population change from *Unbound Prometheus* by David S. Landes.

2  *shot up twenty-five-fold*: Data from "Cotton Textiles and the Great Divergence: Lancashire, India and Shifting Competitive Advantage, 1600–1850," Table 2, Stephen Broadberry and Bishnupriya Gupta, Department of Economics, University of Warwick (2005).

2  *He considered colonialism to be unprofitable*: See "Un Impérialiste Libéral? Jean-Baptiste Say on Colonies and the Extra-European World" by Anna Plassart, *French Historical Studies* 32 (2) (2009): 223–50.

2  *"the number of steam engines has multiplied prodigiously"*: De l'Angleterre et des Anglais by Jean-Baptiste Say.

2  *two acres to feed a horse for a year*: The British Industrial Revolution, 1760–1860 by Gregory Clark, UC Davis.

2  *over three hundred yards deep*: Durham Mining Museum, information relating to Hetton Colliery.

2   *production of the metal soared ninefold*: Data from *British Industrial Revolution* by Clark, and from *Unbound Prometheus* by Landes.

4   *prodigious amounts of coal*: Estimates of Newcomen engine efficiency from *Modern Engineering Thermodynamics* by Robert T. Balmer. The best Newcomen engines had efficiencies of just over 1 percent.

4   *Britain's mines produced 16 million tons every year*: Data from *Unbound Prometheus* by Landes and *A Short History of the British Industrial Revolution* by Emma Griffin.

4   *coal often sold at less than ten shillings per ton*: Data from "Unravelling the Duty: Lean's Engine Reporter and Cornish Steam Engineering" by Alessandro Nuvolari and Bart Verspagen, Eindhoven Centre for Innovation Studies, Netherlands (2005).

4   *James Watt had patented a modification to the Newcomen engine*: For more on the way Watt improved the steam engine, see the next chapter.

4   *roughly quadrupled their duty*: A precise figure comparing the efficiency of Watt and Newcomen engines is difficult to state because the engines varied so much. I've gone for a fourfold improvement based on *Modern Engineering Thermodynamics* by Balmer and *Transactions of the Institution of Civil Engineers* 3 (1) (January 1842).

4   *braking British innovation for thirty years*: For a discussion of how Watt and Boulton used the patent system, see *Against Intellectual Monopoly* by Michele Boldrin and David K. Levine.

5   *the people of England had a love-hate relationship with science*: For details of public attitudes to science see *When Physics Became King* by Iwan Rhys Morus.

5   *England's two universities*: In 1800, there were two universities in England, namely Oxford and Cambridge. There were five in Scotland: St Andrews, Glasgow, Aberdeen, Edinburgh, and Marischal College.

5   *suspicious of novel mathematical techniques*: See *When Physics Became King* by Morus.

5   *condemned as unpatriotic in the conservative* Anti-Jacobin Review: "Mathematics and Meritocracy: The Emergence of the Cambridge Mathematical Tripos" by John Gascoigne, *Social Studies of Science* 14 (4) (1984).

5   *the price never dropped below twenty-eight shillings per ton*: "The Theory and Practice of Steam Engineering in Britain and France, 1800–1850" by Alessandro Nuvolari, *Documents pour l'histoire des techniques* (2010).

6   *The National Conservatory of Arts and Crafts*: Conservatoire national des arts et métiers; see next chapter.

6   *science and mathematics were weapons in a war against superstition*: See *When Physics Became King* by Morus.

### Chapter Two: The Motive Power of Fire

7   *It is necessary*: From *Reflections on the Motive Power of Heat* by Sadi Carnot, edited by R. H. Thurston.

7   *The young man is extremely gentle*: Quoted in Robert Fox's introduction to *Reflections on the Motive Power* by Carnot.

7   *Sadi Carnot was born*: Many biographical details are found in Hippolyte Carnot's memoir of his brother, Sadi.

7   *Lazare, was a gifted mathematician and engineer*: See "Lazare and Sadi Carnot: A Scientific and Filial Relationship" by Charles Coulston Gillispie and Raffaele Pisano, *Volume 19, History of Mechanism and Machine Science*.

8   *the Polytechnic School*: See "The 'École Polytechnique,' 1794–1850: Differences over Educational Purpose and Teaching Practice" by Ivor Grattan-Guiness in the *American Mathematical Monthly*.

8   *Conservatory of Arts and Crafts*: Sadi Carnot's time at this institution, the Conservatoire national des arts et métiers, is discussed in Robert Fox's introduction to his translation of *Reflections on the Motive Power*.

10  *Thompson and Mme Lavoisier's marriage was short*: See *International Women in Science: A Biographical Dictionary to 1950* by Catharine M. C. Haines and Helen M. Stevens.

11  *Clément's lectures*: These appear in *Le Producteur: Journal De L'Industrie, des Sciences et des Beaux Arts*, 1825 issue.

11  *Lazare imagined an ideal mill*: See Robert Fox's introduction to *Reflections on the Motive Power*.

11  *expatriate English engineer*: Samuel Aston built a steam engine known as Wasserkunst Magdeburg in Magdeburg in 1818. See *Grace's Guide to British Industrial History*.

13  *Two aspects greatly interested the Frenchman*: See Carnot's text and *From Watt to Clausius: The Rise of Thermodynamics in the Early Industrial Age*. Author D. S. L. Cardwell says, "By his invention of the expansive principle (1769) this meticulous Scotsman foreshadowed the progressive improvement of heat-engines and the postulation by Sadi Carnot of a general theory of motive power of heat."

15  *This arrangement will run forever*: In this chapter I'm sticking with Carnot's assumption that the same amount of heat enters the engine as leaves it. This assumption isn't right because in fact some heat becomes work. Nonetheless Carnot's logic is valid, and we'll see in later chapters why his conclusions are correct despite this flawed assumption.

16  *The point is that this furnace will never lose any heat*: All the heat that leaves the furnace to drive the forward engine is replenished by the reverse engine.

17  *How much better, though?*: The figures I use are to demonstrate Carnot's arguments and aren't important in themselves.

19  *that all comes from the heat flow*: See appendix I for the details of how Carnot imagined the perfect heat engine worked.

19  *Carnot reckoned, a little over 160°C*: These figures come from Carnot's text.

20  *Rudolf Diesel, who published his theories*: In his book *Theory and Construction of a Rational Heat Motor*, Diesel says, "We may therefore reasonably hope that the new motor will give the economy of combustible claimed for it, as it is based on the principles of the perfect Carnot cycle."

20  *he self-published at a cost of 459.99 francs*: From Robert Fox's introduction to his translation of *Reflections on the Motive Power*.

21  *Our last glimpse is of Sadi Carnot*: See article in *Revue d'Histoire des Sciences* 27 (4) (1974).

### Chapter Three: The Creator's Fiat

23  *I have neither propelled vessels*: Quoted in *James Joule: A Biography* by D. S. L. Cardwell

23  *On May 24, 1842, two brothers in their twenties*: For more biographical information on Joule, see *James Joule: A Biography* by Cardwell and *James Prescott Joule* by Osborne Reynolds.

23   *the population roughly doubled*: National Census and Registrar General's Mid-Year
     Population Estimates, Office for National Statistics.
24   *"Electrical Euphoria" swept the Western world*: For more on this, see "Science and
     Technology: The Work of James Prescott Joule," *Technology and Culture* 17 (4), by
     D. S. L. Cardwell (1976).
24   *Joule was building batteries, electromagnets, and motors*: For details on all of Joule's
     experiments, see *Scientific Papers* by James Joule.
25   *he sent a paper describing it to Britain's most prestigious scientific publication*: See
     J. Young, "Heat, Work and Subtle Fluids: A Commentary on Joule (1850) 'On the
     Mechanical Equivalent of Heat,'" *Philosophical Transactions of the Royal Society A*
     373 (2015): 20140348.
26   *He designed a hand-cranked dynamo with a clever modification*: For more details see
     *Scientific Papers* by Joule. In this experiment Joule used the crank so the water-filled
     tube containing the coil spun in a magnetic field generated by stationary electromag-
     nets. These were powered by batteries.
26   *If he rotated the dynamo in one direction*: Joule's dynamo had a commutator, so it gen-
     erated pulses of current that flowed in the same direction. It didn't generate alternat-
     ing current.
28   *"essentially a holy undertaking"*: From notes Joule made in preparation for a lecture
     he was due to give at the British Association for the Advancement of Science (1873).
28   *"The grand agents of nature are"*: "On the Calorific Effects of Magneto-Electricity,
     and on the Mechanical Value of Heat" by J. P. Joule (1843).
28   *The word* scientist *was mooted at their first meetings*: As reported by William Whewell
     in *Quarterly Review* 51 (1834).
28   *"the subject did not excite"*: A note from Joule's *Collected Papers Vol. 2* (1885).
29   *Also, he was not charismatic* (and subsequent quotes): See *James Prescott Joule* by
     Reynolds.
30   *there was a scheduling foul-up on the day of his talk*: See "Chemistry and the Con-
     servation of Energy: The Work of James Prescott Joule" by John Forrester, *Studies in
     History and Philosophy of Science* 6 (4) (1975).
31   *creating "a lively interest in the new theory"*: From a letter by James Joule written in
     1885.
31   *"I felt strongly impelled at first to rise"*: From a letter by William Thomson written in
     1882.
31   *embellished the story of the Oxford encounter*: Quoted in *Lord Kelvin: An Account of
     His Scientific Life and Work* by Andrew Gray.

## Chapter Four: The Valley of the Clyde

33   *overemphasizing the* r *in Carnot*: From "On the Dissipation of Energy" by William
     Thomson in *Fortnightly Review*, March 1892.
33   *Thomson was in Paris undergoing*: For details of Thomson's life and career see *Energy
     and Empire: A Biographical Study of Lord Kelvin* by Crosbie Smith and M. Norton
     Wise. Also *Degrees Kelvin* by David Lindley and *Lord Kelvin: An Account of His Sci-
     entific Life and Work* by Andrew Gray.
34   *Thomson published a detailed defense*: See *A History of the University of Cambridge,
     Vol. 3, 1750–1870*.

35  *"without doubt the Alma Mater of my scientific youth"*: *Comptes Rendus de l'Académie des Sciences* 121(1895): 582.

35  *"He talks about it ceaselessly all day"*: Letter from Hermann Helmholtz to Frau Helmholtz, 1863.

35  *formidable scientific double act*: For more on James Thomson's influence on his brother William, see *Energy and Empire* by Smith and Wise.

35  *"It is really quite comic"*: Letter from Helmholtz to Helmholtz, 1863.

35  *"a very beautiful piece of reasoning"*: Letter from James to William Thomson, February 22, 1846.

36  *a new vessel slid out of the yards along Glasgow's river Clyde every ten days*: Estimate from records collected by Dr. Strang using "returns furnished to him by various ship-builders and engineers in Glasgow, Dumbarton, Greenock and Port-Glasgow." Quoted in *Biographical Dictionary of Eminent Scotsmen*, article on Henry Bell.

36  *four hundred passengers across the Atlantic*: Many Glasgow-built ships are described in the *Transactions of the Glasgow Archaeological Society*. These journeys didn't always end well. The *City of Glasgow* was lost at sea in 1854. All on board were lost.

36  *Glasgow's population rose*: Reports of medical officer of health, Glasgow as shown in https://www.understandingglasgow.com/indicators/population/trends/historic_population_trend.

36  *"six o'clock had struck by hearing"*: "Memoir of Norman Macleod, D.D." by the Reverend Donald Macleod in the *Christian's Penny Magazine, and Friend of the People* (1876).

36  *"An Account of Carnot's Theory of the Motive Power of Heat"*: *Mathematical and Physical Papers* 1, by Sir William Thomson.

37  *from James, William's older brother, who had devised an ingenious way of testing*: See above.

38  *Of course, such an experiment presented huge practical difficulties*: See above.

39  *Joule's work amplified another doubt Thomson had about the theory*: See above.

## Chapter Five: The Principal Problem of Physics

41  *It is a spectacle for the gods*: Letter from Emil Du Bois-Reymond to Hermann Helmholtz in 1852 (*Dokumente einer Fruendshaft* by Kirsten et al.).

41  *the river Havel opens out into an interconnected system of lakes*: For more on the role of steam in Berlin's parks and nineteenth-century Prussian culture in general, see "Architectures for Steam" by M. Norton Wise (chapter 5 in *The Architecture of Science*).

41  *"one of the most beautiful in Germany"*: All von Moltke's quotes are from an 1851 letter he wrote to his wife (as quoted in Wise's chapter above).

42  *Born in 1821 in Potsdam*: For biographical information on von Helmholtz, see *Helmholtz: A Life in Science* by David Cahan, *Hermann von Helmholtz* by Leo Koenigsberger (translated by Frances A. Welby), and *Hermann Ludwig Ferdinand von Helmholtz* by John Gray McKendrick.

42  *Economically it lagged behind both Britain and France*: For more on the spread of industrialization in nineteenth-century Europe, see *The Unbound Prometheus: Technological Change and Industrial Development in Western Europe from 1750 to the Present* by David S. Landes.

43   *mere 20,000 horsepower*: Data from *The Dictionary of Statistics*, 4th ed., by M. G. Mulhall.

43   *abolished serfdom in 1807*: The 1807 October Edict.

43   *formed a customs union in 1834*: For more on the economic impacts of this, see Wolfgang Keller and Carol Hua Shiue, *The Trade Impact of the Zollverein*, March 2013, CEPR Discussion Paper no. DP9387.

43   *"hostile tax-gatherers and customs house officials"*: From an 1819 petition by economist Friedrich List, head of the Union of Merchants.

43   *Between 1840 and 1860, fixed steam engine capacity increased tenfold*: Data from *Dictionary of Statistics* by Mulhall (reported in *Modern Capitalist Culture* by Leslie A. White).

43   *ten thousand miles of track*: See *Unbound Prometheus* by Landes.

43   *better-funded German education system*: See *The German Genius* by Peter Watson, p. 237.

43   *redefined their purpose*: See "The Growth of Professorial Research in Prussia, 1818 to 1848—Causes and Context" by R. Steven Turner in *Historical Studies in the Physical Sciences* 3 (1971).

43   *A like-minded group coalesced*: See *Helmholtz* by Cahan, p. 72.

43   *vitalism*: In German, this was known as *Lebenskraft* or *life force*.

44   *"Respiration is then a combustion"*: *Memoire sur la Chaleur* (1780) by Antoine Lavoisier and Pierre Simon de Laplace.

44   *With this in mind, two scientists*: For details of these experiments, see article in the *Medical Times* in 1846 by Robert Rigg.

45   *Helmholtz pointed out that the carbohydrate molecules*: For a lucid description of how Helmholtz critiqued vitalism, see *Life's Ratchet: How Molecular Machines Extract Order from Chaos* by Peter M. Hoffmann.

47   *"If perpetual motion is impossible"*: "On the Interaction of Natural Forces," a lecture delivered by Helmholtz on February 7, 1854, in Königsberg.

47   *Imagine a frictionless slope*: This is my thought experiment.

48   *"lay before physicists"*: "On the Conservation of Force" by H. Helmholtz, read before the Physical Society of Berlin (1847).

49   *"Whether by the development"*: See above.

## Chapter Six: The Flow of Heat and the End of Time

51   *Over and above his direct contributions to science*: "Death of Professor Magnus," *Nature* 1 (1870): 607, by John Tyndall.

51   *"reminiscent of English"*: See entry on Magnus in *Allgemeine Deutsche Biographie* by A. W. Hofmann.

51   *Magnus brought it before the colloquium*: See "Rudolph Clausius: A Pioneer of the Modern Theory of Heat" by Stefan L. Wolff, *Vacuum* 90 (2013); and "The Berlin School of Thermodynamics Founded by Helmholtz and Clausius" by Werner Ebeling and Dieter Hoffman, *European Journal of Physics* 12 (1991).

52   *The sixth son of a Lutheran pastor*: Biographical information on Clausius is scant. The basic facts are in "Obituary Notices of Fellows Deceased," *Proceedings of the Royal Society of London* 48:i–xxi. Also see *Great Physicists* by William H. Cropper.

52   *the father of theoretical physics*: See *The Second Physicist: On the History of Theoretical Physics in Germany* by Christa Jungnickel and Russell McCormmach.

52 *Clausius described his breakthrough in a historic paper*: "On the Moving Force of Heat and the Laws Which Can Be Deduced Therefrom" by Rudolf Clausius, *Annalen der Physik*, 1850.

52 *With flawless logic, Clausius reasoned as follows*: See appendix 11 for more details.

55 *This resembles a modern refrigerator*: For more on how refrigerators work, see chapter 10.

56 *For illustration, the ideal engine*: The figures I use are mine, not Clausius's.

58 *reading it in Glasgow*: See *Energy and Empire: A Biographical Study of Lord Kelvin* by Crosbie Smith and M. Norton Wise.

59 *a paper Thomson published in*: "On a Universal Tendency in Nature to the Dissipation of Mechanical Energy" by Sir William Thomson (Lord Kelvin), *Proceedings of the Royal Society of Edinburgh for April 19, 1852.*

59 *one hundred thousand destitute refugees*: Glasgow City Council News Archive, June 2018.

59 *Thomson's younger brother John contracted typhus*: See *Energy and Empire* by Smith and Wise.

60 *"I believe that no physical action"*: Quoted in *The Science of Energy* by Crosbie Smith and *Energy and Empire* by Smith and Wise.

61 *"we must admire the sagacity of Thomson"*: "On the Interaction of Natural Forces" by Helmholtz.

## Chapter Seven: Entropy

63 *Die Entropie*: From *Annalen der Physik and Chemie*, 1865.

63 *mercury expands by around 0.018 percent*: I have used a value for the volume thermal expansion coefficient of mercury of 0.00018/K taken from www.EngineeringTool Box.com.

63 *Which substance should you trust?* I have chosen an extreme scenario to demonstrate the unreliability of using the thermal properties of substances as a way of measuring temperature. Water just above freezing behaves unlike most other liquids. For many common purposes, the conveniences of this method more than compensate for any inconsistencies.

64 *generates in units called PUFFs*: For the purposes of this illustration, the value of a PUFF is unimportant. It could be, say, the work required to raise a weight of one kilogram by one meter.

65 *This engine can operate as a thermometer*: For a more technical explanation see Richard Feynman's lecture "The Laws of Thermodynamics," https://www.feynmanlectures.caltech.edu/I_44.html.

66 *Clausius remained hard at work*: Clausius was the first chair in physics at the Eidgenössische Technische Hochschule, Zurich, which was founded in 1855. Albert Einstein was a student there from 1896 to 1900.

67 *Clausius conceived of entropy to capture mathematically*: See *The Mechanical Theory of Heat, with Its Applications to the Steam Engine and to the Physical Properties of Bodies* by Rudolf Clausius.

67 *When a quantity of heat flows out of a hot room*: In my example I'm using rooms, but this applies to any object or body. Clausius is interested in the heat that flows in and out of the cylinders in steam engines.

69 *sunshine, or wind*: The ultimate source of wind power is heat from the sun. This

causes uneven heating of the earth's atmosphere and surface, which in turn causes winds.

69  *he added his own coinage,* entropy: In the "Ninth Memoir" in his *Mechanical Theory of Heat,* Clausius writes, "I propose to call the magnitude S the entropy of the body, from the Greek word τροπή transformation. I have intentionally formed the word *entropy* so as to be as similar as possible to the word *energy*; for the two magnitudes to be denoted by these words are so nearly allied in their physical meanings that a certain similarity in designation appears to be desirable."

69  *The laws state:* These are the final lines in the "Ninth Memoir" of Clausius's *Mechanical Theory of Heat.*

70  *"One species does change into another":* Darwin's "Red Notebook," p. 130.

70  *The book espoused the idea of uniformitarianism: Principles of Geology: Being an Attempt to Explain the Former Changes of the Earth's Surface, by Reference to Causes Now in Operation* by Charles Lyell (1830–33).

71  *he had become convinced the doctrine of uniformitarianism was flawed:* See *Lord Kelvin and the Age of the Earth* by Joe D. Burchfield.

71  Macmillan's Magazine: "On the Age of the Sun's Heat" by Sir William Thomson (Lord Kelvin), *Macmillan's Magazine* 5 (March 5, 1862).

71  *"Essential principles of thermodynamics have been overlooked by geologists":* From "On the Secular Cooling of the Earth" by William Thomson, *Transactions of the Royal Society of Edinburgh* (1862).

72  *"I am greatly troubled at the short duration": More Letters of Charles Darwin* edited by F. Darwin and A. C. Seward, letter dated July 24, 1869.

72  *"odious spectre": Letters of Wallace* edited by J. Marchant (1916), letter from Darwin to Wallace dated January 26, 1870.

72  *is "one of the gravest as yet advanced": On the Origin of Species* by Charles Darwin.

72  *that many philosophers are not as yet willing to admit:* See above.

73  *"allows the time claimed by the geologist and biologist for the process of evolution": Radiation and Emanation* by Ernest Rutherford (1904).

## Chapter Eight: The Motion We Call Heat

75  *he was at pains to point out:* See "On the Moving Force of Heat and the Laws Which Can Be Deduced Therefrom" by Rudolf Clausius, *Annalen der Physik,* 1850; and "The Nature of the Motion Which We Call Heat" by Rudolf Clausius, *Annalen der Physik,* 1857. Also see "Clausius and Maxwell's Kinetic Theory of Gases" by Elizabeth Wolfe Garber, *Historical Studies in the Physical Sciences* 2 (1970): 299–319; and the introduction to *Kinetic Theory,* vol. 1, by Stephen G. Brush.

75  *clear from his later writings:* See "Nature of the Motion" by Clausius.

75  *to prompt Clausius:* The most direct prompt was an article by Karl August Kroenig, "Grundzüge einer Theorie der Gase," *Annalen der Physik,* 1856. See "Nature of the Motion" by Clausius.

76  *when it was mooted by a Swiss polymath called Daniel Bernoulli: Hydrodynamics, or Commentaries on the Forces and Motions of Fluids* (1738). Chapter 10 of this Latin text is entitled "On the Properties and Motions of Elastic Fluids, Especially Air."

76  *In many ways Bernoulli's idea was too advanced for its time:* See the introduction to *Kinetic Theory* by Brush.

76  *Bernoulli lived in an age when academics*: For more on Bernoulli's life and work see "Bernoulli, Daniel" by Hans Straub from the *Complete Dictionary of Scientific Biography*.

76  *Bernoulli applied these principles to fluids*: *Hydrodynamics* by Bernoulli.

77  *"very small particles in very rapid motion"*: From chapter 10 of Bernoulli's *Hydrodynamics*. There's an English translation in *Kinetic Theory* by Brush.

78  *"are in more violent motion"*: See above.

78  *"The paper is nothing but nonsense," wrote the society's referee*: The referee was the astronomer Sir John William Lubbock. The letter is in the Royal Society Archives.

78  *As a theoretician and a gifted mathematician*: See *The Second Physicist: On the History of Theoretical Physics in Germany* by Christa Jungnickel and Russell McCormmach.

79  *The style of Clausius's paper is remarkable*: "Nature of the Motion" by Clausius.

80  *As the average speed of an oxygen molecule*: Lighter molecules such as hydrogen and helium do leak out of the earth's atmosphere because a proportion of them do move at speeds greater than the earth's escape velocity.

81  *a Dutch meteorologist named Christoph Buys Ballot wrote a paper*: "On the Nature of the Motion Which We Call Heat and Electricity" by C. H. D Buijs-Ballot (Buys Ballot), *Annalen der Physik*, 1858.

81  *a second paper on kinetic theory with a clever*: "On the Mean Lengths of the Paths Described by the Separate Molecules of Gaseous Bodies" by Rudolf Clausius, *Annalen der Physik*, 1858.

82  *English version of Clausius's second paper*: From *London, Edinburgh and Dublin Philosophical Magazine and Journal of Science*, 4th ser., February 1859.

### Chapter Nine: Collisions

83  *a twenty-five-year-old named James Clerk Maxwell*: For more on Maxwell's life and career, see *The Life of James Clerk Maxwell* by Lewis Campbell and *The Man Who Changed Everything: The Life of James Clerk Maxwell* by Basil Mahon.

83  *"second dip of the season" and "gymnastics on a pole afterwards"*: Letter from Maxwell to Miss Cay, February 1857.

84  *Maxwell had written his first scientific paper*: The paper, "On the Description of Oval Curves, and Those Having a Plurality of Foci," was read to the Royal Society of Edinburgh on April 6, 1846.

84  *"indirect and enigmatical"*: "Biographical Outline" in *Life of James Clerk Maxwell* by Campbell.

85  *"necessary to one another"*: Letter from Maxwell to Miss Cay, February 1858.

85  *first love was his cousin Lizzie Cay*: See *Man Who Changed Everything* by Mahon.

85  *"The lady was neither pretty, nor healthy nor agreeable, but much enamored of him"*: As quoted in *Degrees Kelvin* by David Lindley.

85  *"James, it's time you went home"*: As quoted in *Degrees Kelvin* by David Lindley.

85  *"Will you come along with me"*: Poem written by Maxwell in 1858.

86  *In this, Maxwell was inspired by the work of astronomers*: In 1809, the mathematician and physicist Carl Friedrich Gauss published a clever way of determining the orbit of the planetoid Ceres that best fitted astronomers' observations.

86  *by the astronomer John Herschel*: The paper, "Probabilities," appeared in the *Edinburgh Review*, 1850.

86   *"The true logic of this world"*: Maxwell in a letter to his friend Lewis Campbell, 1850.

87   *ask a competent marksman*: Herschel used exactly the example of the aftermath of a rifleman's target practice. He sets the puzzle to find the target when you only have the bullet holes left in the wall. Then he writes, "Suppose the rifle replaced by a telescope duly mounted; the [target] by a star on the concave surface of the heavens, always observed for a succession of days at the same sidereal time; the marks on the wall by the degrees, minutes, and seconds, read off on divided circles; and the marksman by an observer; and we have the case of all direct astronomical observation where the place of a heavenly body is the thing to be determined."

88   *Maxwell started by reiterating*: "Illustrations of the Dynamical Theory of Gases" by James Clerk Maxwell was read at the meeting of the British Association for the Advancement of Science in Aberdeen in 1859 and published in the *Philosophical Magazine* in 1860.

88   *The chances drop off in a similar way*: Molecular speeds do not follow a bell curve exactly. The latter is perfectly symmetrical—the chances of overshoot are equal to that of an undershoot. Molecules in a gas, however, have a theoretical minimum speed—namely zero. But there is no theoretically defined maximum speed although the chances of finding molecules at three or four times the average are vanishingly small. In short, molecular speeds form a bell curve that's slightly stretched toward the faster speeds.

89   *gas's pressure varies*: This is true as long as the gas's temperature remains constant.

90   *"The only experiment I have met with"*: See above.

90   *"I am getting quite fond of it"*: Letter from Maxwell to George Gabriel Stokes, May 1859.

90   *Unfortunately for Maxwell*: John S. Reid, "Maxwell at Aberdeen," in *James Clerk Maxwell: Perspectives on His Life and Work*, ed. Raymond Flood, Mark McCartney, and Andrew Whitaker, 17–42, 304–10.

91   *Here Maxwell returned to the question he had been investigating in Aberdeen*: See "The Bakerian Lecture: On the Viscosity or Internal Friction of Air and Other Gases," received November 23, 1865, read February 8, 1866.

93   *Physicists were quick to spot the significance*: See the introduction to *Kinetic Theory*, vol. 1, by Stephen G. Brush.

93   *"Ho, Maxwell, cannot you get out?"*: Quoted in *Life of James Clerk Maxwell* by Campbell.

93   *his intuition was telling him*: See chapter 17.

## Chapter Ten: Counting the Ways

95   *Mathematics is a language*: As quoted in *Willard Gibbs* by Muriel Rukeyser.

95   *The staccato opening chords of*: From the website of the Vienna Philharmonic Orchestra. The *Eroica* was performed on Friday, June 1, 1866, www.wienerphilharmoniker.at/converts/archive.

95   *a bearded, bespectacled twenty-two-year-old named Ludwig Boltzmann*: For more on Boltzmann's life and career see *Ludwig Boltzmann: The Man Who Trusted Atoms* by Carlo Cercignani and *Boltzmann's Atom: The Great Debate That Launched a Revolution in Physics* by David Lindley.

95   *another man from another continent, Josiah Willard Gibbs*: For more on Gibbs's life

and work, see *Willard Gibbs* by Muriel Rukeyser, *Josiah Willard Gibbs: The History of a Great Mind* by Lynde Phelps Wheeler, and *Biographical Memoir of Josiah Willard Gibbs, 1839–1903* by Charles S. Hastings.

96  *"regional finance commissar"*: See *Ludwig Boltzmann: Leben und Briefe* by Walter Höflechner, as quoted in *Boltzmann's Atom* by Lindley.

96  *"stubby fingers and pudgy hands"*: Reminiscence by Boltzmann's onetime assistant Stefan Meyer. As quoted in *Boltzmann's Atom* by Lindley.

97  *agreed to fund a Physics Institute at Vienna University*: For more on the creation of this department, see *100 Jahre Physik an der Universität Wien* by Wolfgang L. Reiter (2015).

97  *Years later Boltzmann would remember his time here as the happiest*: See *Populäre Schriften* by Ludwig Boltzmann.

97  *Josef Loschmidt, a lecturer twenty-three years older than Boltzmann*: See "Scientific Discussion and Friendship between Loschmidt and Boltzmann" by Boltzmann's grandson Dieter Flamm and "In Memoriam: Ludwig Boltzmann: A Life of Passion" by Wolfgang L. Reiter, *Physics in Perspective* 9 (2007).

97  *"my sweet fat darling"*: From Dieter Flamm's *Life and Personality of Ludwig Boltzmann*, as quoted in *Ludwig Boltzmann* by Cercignani.

97  *Loschmidt had used their ideas to deduce the diameter of a single particle of air*: See "Joseph Loschmidt, Physicist and Chemist" by Alfred Bader and Leonard Parker, *Physics Today* 54 (3) (2001): 45.

97  *"Ever higher surges"*: As quoted in *Boltzmann's Atom* by Lindley.

98  *"Further Studies in the Thermal Equilibrium of Gas Molecules"*: By Ludwig Boltzmann, originally published in *Sitzungsberichte der Akademie der Wissenschaften* (Vienna).

99  *To simplify his analysis, he employed a mathematical trick*: See above paper.

104  *"not so accessible"*: From *Ludwig Boltzmann* by Höflechner, as quoted in *Boltzmann's Atom* by Lindley.

104  *Boltzmann found them unwelcoming*: See *Ludwig Boltzmann* by Cercignani.

104  *Gibbs came from an intellectual tradition*: See *Willard Gibbs* by Muriel Rukeyser.

105  *Willard Gibbs Sr. determined to solve this problem*: See above.

105  *The North's eventual victory*: See *How the Railroads Won the War*, Smithsonian American Art Museum, February 2015.

106  *"The notion of entropy . . . will doubtless seem to many far-fetched"*: From "Graphical Methods in the Thermodynamics of Fluids" by Josiah Willard Gibbs, *Transactions of the Connecticut Academy*, 1873.

106  *In 1873, in his first two scientific papers*: See above paper and "A Method of Geometrical Representation of the Thermodynamic Properties of Substances by Means of Surfaces" by Josiah Willard Gibbs, *Transactions of the Connecticut Academy*, 1873.

107  *Darwin noted this effect*: "Hence the potatoes, after remaining for some hours in the boiling water were not cooked. I found out this, by overhearing my two companions discussing the cause; they had come to the simple conclusion, 'that the cursed pot (which was a new one) did not choose to boil potatoes.'" Darwin, *The Voyage of the Beagle*.

109  *much of the world's electricity*: See data from the International Energy Agency at iea .org.

109  *so-called phase changes*: For a more technical description see *Finn's Thermal Physics* by Andrew Rex.

109 *Archaeology suggests*: See for instance "Microstratigraphic Evidence of In Situ Fire in the Acheulean Strata of Wonderwerk Cave, Northern Cape Province, South Africa" by Francesco Berna, Paul Goldberg, Liora Kolska Horwitz, James Brink, Sharon Holt, Marion Bamford, and Michael Chazan, *Proceedings of the National Academy of Sciences*, April 2, 2012.

110 *Caribbean, Europe, and even India*: For more on Tudor and the ice trade, see *Refrigeration: A History* by Carroll Gantz.

110 *estimated ninety thousand people*: See *The Ice Crop: How to Harvest, Store, Ship and Use Ice* by Theron Hiles.

110 *Norway exported a million tons of ice*: See *Melting Markets: The Rise and Decline of the Anglo-Norwegian Ice Trade, 1850–1920* by Bodil Bjerkvik Blain, Working Papers of the Global Economic History Network (GEHN) no. 20/06 (2006).

110 *But could ice be made artificially?*: For details on how this developed and more on people such as James Harrison and Carl Linde, see *Refrigeration* by Gantz.

111 *German scientist, engineer, and entrepreneur named Carl Linde*: See "Carl Von Linde: A Pioneer of 'Deep' Refrigeration" by J. H. Awbery, *Nature*, 1942.

111 *By 1875, he was using thermodynamic charts*: See *The Development of Science and Technology in Nineteenth-Century Britain: The Importance of Manchester* by Donald Cardwell.

112 *its pressure drops and it cools*: Typically as the coolant leaves the expansion valve, it's a mix of liquid and gas.

113 *"Yet we all believed what Gibbs wrote"*: "How the Works of Professor Willard Gibbs Were Published" by A. E. Verrill, *Science* 61 (1925): 41–42.

## Chapter Eleven: "The Terroristic Nimbus"

115 *I am conscious of being only*: *Lectures on Gas Theory* by Ludwig Boltzmann, translated by Stephen G. Brush.

115 *he met the nineteen-year-old Henriette von Aigentler*: *Ludwig Boltzmann: The Man Who Trusted Atoms* by Carlo Cercignani and *Boltzmann's Atom: The Great Debate That Launched a Revolution in Physics* by David Lindley.

115 *"a comrade in a shared endeavor"*: From *Ludwig Boltzmann–Henriette von Aigentler Briefwechsel*, edited by Dieter Flamm, as quoted in *Boltzmann's Atom* by Lindley.

116 *"Unfortunately, I have right now very little time to read or study"*: See above.

116 *"terroristic nimbus cloud"*: From the paper "Ueber den Zustand des Waermegleichgewichtes eines Systems von Koerpern mit Ruecksicht auf die Schwerkraft" by Joseph Loschmidt (1876).

117 *he wrote two papers, published in 1877*: "On the Relation of a General Mechanical Theorem to the Second Law of Thermodynamics" and "On the Relationship between the Second Fundamental Theorem of the Mechanical Theory of Heat and Probability Calculations Regarding the Conditions for Thermal Equilibrium" by Ludwig Boltzmann.

117 *"Elegance is for the tailor"*: Einstein quotes Boltzmann as having said this in his book *On the Special and the General Relativity Theory: A Popular Exposition*.

117 *the following equation*: From the translation of Boltzmann's paper "On the Relationship between the Second Fundamental Theorem" by Kim Sharp and Franz Matschinsky, *Entropy*, 2015, 17.

118  *magnum opus, a 371-page epic paper*: "On the Equilibrium of Heterogeneous Substances" by Josiah Willard Gibbs, *Transactions of the Connecticut Academy*.

118  *First law: The energy*: Gibbs quoted these laws in Rudolf Clausius's original German, "Die Energie der Welt ist constant" and "Die Entropie der Welt strebt einem Maximum zu."

118  *turning the two laws into one new law*: Gibbs's work is rigorous and complex, and the ideas in this paper are an extension of principles he elucidated in his 1873 papers. My aim is to give a feel for his ideas and convey their scope.

121  *Gibbs free energy*: More technically, Gibbs free energy is the energy available at constant pressure and temperature. This has wide applicability because many chemical reactions do occur under these conditions, especially in biochemistry. For an excellent and not too technical explanation see the Khan Academy video and website https://www.khanacademy.org/science/biology/energy-and-enzymes/free-energy -tutorial/a/gibbs-free-energy. There is another term, *Helmholtz free energy*, which deals with processes that occur in systems at constant volume and temperature.

121  *very first step in the process—photosynthesis*: An excellent primer on all the steps in this process are on the Khan Academy website.

123  *sunshine and a sewer*: An excellent, pithy video by physicist Sean Carroll on the YouTube channel minutephysics describes this: "What Is the Purpose of Life?," Big Picture ep. 5/5. As Carroll puts it, "Living organisms ultimately depend on and facilitate the universe's tendency to increase entropy." Also, see "Entropy in Biology" by Jayant B. Udgaonkar in *Resonance: Journal of Science Education*, 2001.

124  *already fragile psyche*: See *Ludwig Boltzmann* by Cercignani, *Boltzmann's Atom* by Lindley, and "In Memoriam: Ludwig Boltzmann: A Life of Passion" by Wolfgang L. Reiter, *Physics in Perspective* 9 (2007).

124  *phenomenalism had become his passion*: For a summary of Mach's point of view, see *Intellectual Mastery of Nature: Theoretical Physics from Ohm to Einstein*, vol. 2, by Christa Jungnickel and Russell McCormmach. For more details, see *The Science of Mechanics* by Ernst Mach (1883).

125  *exhausting debates*: See, for example, "Helm and Boltzmann: Energetics at the Lübeck Naturforscherversammlung" by Robert Deltete, *Synthese* 119 (1999).

125  *they turned to the work of Josiah Willard Gibbs*: See above paper and "Gibbs and the Energeticists" by Robert Deltete in Boston Studies in the Philosophy of Science book series (BSPS, vol. 167) (1995).

125  *young physics lecturer in Munich named Max Planck*: For biographical information on Planck, see www.nobelprize.org/prizes/physics/1918/planck/biographical/.

125  *"The second law of the mechanical theory"*: From *Vaporization, Melting and Sublimation* by Max Planck (1882). Quoted in *Black-Body Theory and the Quantum Discontinuity, 1894–1912* by Thomas S. Kuhn.

125  *"I don't believe that atoms exist!"*: As quoted in *Boltzmann's Atom* by Lindley.

125  *"ran around in my head"*: See above.

126  *"Whether I will soon be alone"*: *Ludwig Boltzmann: Life and Letters* by Walter Höflechner, as quoted in *Boltzmann's Atom* by Lindley.

126  *"Shouldn't the irresistible urge"*: See above.

126  *Zermelo's papers*: *Wissenschaftliche Abhandlung von Ludwig Boltzmann* edited by Fritz Hasenhoehrl, as quoted in *Boltzmann's Atom* by Lindley.

## Chapter Twelve: Boltzmann Brains

127    *I sleep badly*: *Ludwig Boltzmann: Life and Letters* by Walter Höflechner, as quoted in *Boltzmann's Atom: The Great Debate That Launched a Revolution in Physics* by David Lindley.

127    *universe must have been born*: See "On Zermelo's Paper 'On the Mechanical Explanation of Irreversible Processes'" by Ludwig Boltzmann, 1897.

127    *statistically unlikely low-entropy state*: For a detailed discussion of Boltzmann's response to Zermelo, see *From Eternity to Here: The Quest for the Ultimate Theory of Time* by Sean Carroll.

128    *"A living being"*: See "On Zermelo's Paper" by Boltzmann, from *Kinetic Theory*, vol. 2, by Stephen G. Brush.

128    *This kind of reasoning is known as the anthropic principle*: See *From Eternity to Here* by Carroll.

130    *Boltzmann brain*: See "Can the Universe Afford Inflation?" by Andreas Albrecht and Lorenzo Sorbo, *Physical Review D* 70 (2004); and see *From Eternity to Here* by Carroll.

130    *"For some reason, the universe"*: *The Feynman Lectures on Physics* 1, chap. 46, "Ratchet and Pawl."

130    *"Papa sweats and swears all the time"*: From "Ludwig Boltzmann and His Family" by Ilse M. Fasol-Boltzmann (Boltzmann's granddaughter) in the introduction to *Ludwig Boltzmann Principien der Naturfilosofi: Lectures on Natural Philosophy, 1903–1906*.

130    *"Just when I received"*: *Ludwig Boltzmann* by Höflechner, as quoted in *Boltzmann's Atom* by Lindley.

131    *"Certainly, one is building on"*: From *Elementary Principles in Statistical Mechanics Developed with Special Reference to Rational Foundations of Thermodynamics* by Josiah Willard Gibbs.

131    *"Journey of a German Professor to Eldorado"*: An English translation of this essay is printed in *Ludwig Boltzmann: The Man Who Trusted Atoms* by Carlo Cercignani.

132    *"The lecture was really"*: "Looking Back" by Lise Meitner, *Bulletin of Atomic Scientists* 20 (1954).

132    *When Boltzmann's daughter*: See "In Memoriam: Ludwig Boltzmann: A Life of Passion" by Wolfgang L. Reiter, *Physics in Perspective* 9 (2007).

## Chapter Thirteen: Quanta

133    *In 1900, Max Planck*: "On and Improvement of Wien's Equation for the Spectrum," October 1900, and "On the Theory of the Energy Distribution Law of the Normal Spectrum," December 1900, by Max Planck, translated by D. ter Haar and Stephen G. Brush.

134    *Scientists' understanding of radiating heat*: The first physicist to try to analyze the way objects radiate heat and light was Gustav Kirchhoff (1824–87). He was followed by Ludwig Boltzmann and Wilhelm Wien. See *Black-Body Theory and the Quantum Discontinuity, 1894–1912* by Thomas S. Kuhn, and *The Bumpy Road: Max Planck from Radiation Theory to the Quantum, 1896–1906* by Massimiliano Badino.

134  *empty space is filled with taut "strings" of this kind*: Not to be confused with the modern concept *string theory*.

135  *Imagine along its length, there are tiny compass needles*: Maxwell's argument is symmetric. Instead of an oscillating electron, you could have an oscillating magnet. In which case, the wiggles would travel down magnetic fields, creating an electric field as it does so. In this case, the magnetic force would serve as the "tension" and the electric field as the "weight."

136  *a potter's kiln*: The way kiln-like objects radiate light is known as black body radiation. The term is used because such objects absorb all the light radiation falling on them.

137  *he was attracted to absolute laws such as the first law of thermodynamics*: Planck's doctoral thesis was on thermodynamics.

137  *He disliked Boltzmann's probabilistic explanation*: In a letter to his friend Leo Graetz, Planck wrote, "To maintain that change in nature always proceeds from [states of] lower to higher probability would be totally without foundation." See *Black-Body Theory* by Kuhn and *Bumpy Road* by Badino. (The latter describes the scientific dispute between Boltzmann and Planck in the 1890s.) See also *The Odd Couple: Boltzmann, Planck and the Application of Statistics to Physics, 1900–1913* by Massimiliano Badino (2009).

138  *friend of Planck's named Heinrich Rubens*: For details on the work done at Imperial Physical Technical Institute, see "Revisiting the Quantum Discontinuity" by Jochen Büttner, Olivier Darrigol, Dieter Hoffmann, Jürgen Renn, and Matthias Schemmel for the Max Planck Institute für Wissenschaftsgeschichte (2000).

138  *the English physicist Lord Rayleigh*: This is now commonly referred to as the Rayleigh-Jeans law.

139  *79 different wavelengths*: For details of this calculation, see the article "Black Body Radiation" by J. Oliver Linton.

140  *"an act of despair"*: Letter from Planck in 1931 to Robert Williams Wood.

140  *"probability considerations into the electromagnetic theory"*: "Theory of the Energy Distribution Law" by Planck.

142  *the birth of quantum physics*: Planck first used the term a year later in his paper "Ueber die Elementarquanta der Materie und der Elektricitaet." But the concept appears in his 1900 paper.

142  *"It gave me particular satisfaction"*: "The Origin and Development of the Quantum Theory" by Max Planck, translated by H. T. Clarke and L. Silberstein. The Nobel Prize address delivered before the Royal Swedish Academy of Sciences at Stockholm, June 2, 1920.

## Chapter Fourteen: Sugar and Pollen

143  *Boltzmann is magnificent*: Letter from Einstein to Mileva Maric, September 13, 1900.

143  *"I am conscious of being only"*: *Lectures on Gas Theory* by Ludwig Boltzmann, translated by Stephen G. Brush.

143  *Einstein's science springs*: There are many biographies of Einstein, such as *Einstein: His Life and Universe* by Walter Isaacson and *Subtle Is the Lord: The Science and the Life of Albert Einstein* by Abraham Pais.

143   *his "miracle year"*: The four papers Einstein published in 1905 are "On a Heuristic Viewpoint concerning the Production and Transformation of Light," "On the Motion of Small Particles Suspended in a Stationary Liquid, as Required by the Molecular Kinetic Theory of Heat," "On the Electrodynamics of Moving Bodies," and "Does the Inertia of a Body Depend upon Its Energy Content?"

144   *"I will soon have graced"*: Letter from Einstein to Mileva Maric, April 4, 1901.

144   *"I'll be mad with joy"*: Letter from Einstein to Mileva Maric, April 15, 1901.

144   *"I was able to do a full day's work"*: From *The Private Albert Einstein* by Peter Bucky and Allen G. Weakland.

145   *Einstein's fascination with thermodynamics*: See "Thermodynamics in Einstein's Thought" by Martin J. Klein, *Science* 157 (1967); "Einstein before 1905: The Early Papers on Statistical Mechanics" by Clayton A. Gearhart, *American Journal of Physics*, 1990; "Insuperable Difficulties: Einstein's Statistical Road to Molecular Physics" by Jos Uffink, Institute for History and Foundations of Science, Utrecht University; and "Einstein's Approach to Statistical Mechanics: The 1902–04 Papers" by Luca Peliti and Raul Rechtman, *Journal of Statistical Physics* (2016).

145   *Einstein regarded as revolutionary*: In May 1905, Einstein wrote to his friend Conrad Habicht, "I promise you four papers The first deals with radiation and the energetic properties of light and is very revolutionary."

145   *"that the energy of light"*: "On a Heuristic Viewpoint" by Einstein.

145   *"interpreted in accordance"*: From the above paper.

146   *"readily understood"*: "other related phenomena connected with the emission or transformation of light are more readily understood": From the above paper.

146   *"Interpretation of the Expression"*: See section 6 in the above paper.

147   *he completed his dissertation*: Its title is "A New Determination of Molecular Dimensions."

148   *Using their numbers, he estimated*: Einstein also deduced that some water molecules stick to the sugar molecules, increasing their effective size.

148   *This time he settled on Brownian motion*: See "On the Motion of Small Particles" by Einstein, and "Einstein and the Existence of Atoms" by Jeremy Bernstein, *American Journal of Physics* 74 (10) (October 2006).

150   *"Let us hope that"*: These are last words in "On the Motion of Small Particles" by Einstein.

150   *In just four years*: A German researcher named Henry Siedentopf wrote to Einstein a few months after the latter's paper appeared saying observations from a powerful new microscope seemed to confirm his hypothesis. But Jean Perrin's meticulous and rigorous work is regarded by historians as the evidence that put Einstein's conclusions beyond doubt. Perrin received the Nobel Prize in 1926 for "having put definite end to the long struggle regarding the real existence of molecules."

150   *Perrin's attention to detail was impressive*: For details see "Mouvement Brownien et Molécules" by Jean Perrin, *Journal de Physique Théorique et Appliquée*, April 15, 1909; "Evident Atoms: Visuality in Jean Perrin's Brownian Motion Research" by Charlotte Bigg, *Studies in History and Philosophy of Science* 39 (2008); and "An Experiment to Measure Avogadro's Constant. Repeating Jean Perrin's Confirmation of Einstein's Brownian Motion Equation" by Lew Brubacher, *Chem 13 News*, May 2006.

151   *the distance the particles drifted*: Perrin used other methods to test the existence of atoms and molecules. For instance, he measured how particles suspended in a verti-

cal column of liquid gradually sank under gravity. Comparing the concentrations of particles at different heights provided Perrin with further evidence.

152 *"In my old age I can accept relativity"*: Quoted in "Einstein and the Existence of Atoms" by Bernstein.

154 *"We might say that the principle"*: From the 1946 essay "E = MC², " which was part of the collection *The Theory of Relativity and Other Essays* by Albert Einstein.

## Chapter Fifteen: Symmetry

155 *Gentlemen: I do not*: As quoted in *Hilbert-Courant* by Constance Reid.

155 *Before Noether's theorem*: Quoted in Nathan Jacobsen's introduction to Emmy Noether's collected works.

155 *Emmy Noether was born*: For biographical information on Noether, see *Emmy Noether, 1882–1935* by Auguste Dick, translated by H. I. Blocher; *Symmetry and the Beautiful Universe* by Leon M. Lederman and Christopher T. Hill; and *Emmy Noether's Wonderful Theorem* by Dwight E. Neuenschwander.

155 *"Although a woman"*: As quoted in *Emmy Noether* by Dick.

156 *"This winter I'm giving a course"*: Letter from Emmy Noether to H. Hasse, see above.

158 *contain a symmetry*: Noether's theorem applies to "continuous symmetry" but not to "discrete symmetry." In other words, it applies to systems that change gradually such as the rotation of a circle, but not to systems that change in steps such as when a square is turned.

159 *And that means in turn that energy is not conserved*: See the following article by cosmologist Sean Carroll: https://www.preposterousuniverse.com/blog/2010/02/22/energy-is-not-conserved/.

160 *crux of Einstein's long intellectual battle with the great Danish physicist Niels Bohr*: For details about this, see *Quantum: Einstein, Bohr and the Great Debate about the Nature of Reality* by Manjit Kumar.

160 *"He does not play dice with the universe"*: From a letter from Einstein to Max Born in 1926.

162 *Einstein's first partner*: See *Einstein's Berlin: In the Footsteps of a Genius* by Dieter Hoffmann.

162 *design, patent, and market a refrigerator*: See "The Einstein-Szilard Refrigerators" by Gene Dannen, *Scientific American*, January 1997.

162 *former student named Leo Szilard*: See *Genius in the Shadows* by William Lanouette with Bela Silard.

164 *"The invention never did get on the market"*: As quoted in *Einstein's Berlin* by Hoffmann.

165 *"howled like a jackal"* and *"wailed like a banshee"*: As quoted in "Einstein-Szilard Refrigerators" by Dannen.

165 *"From week to week"*: Letter from Szilard to Einstein, September 27, 1930.

166 *"Suffered?" Hilbert replied*: From *Hilbert-Courant* by Reid.

## Chapter Sixteen: Information Is Physical

167 *Every time you search the internet*: By this I don't mean warmth due to carbon dioxide release. I'm referring to waste heat from data centers.

167　*a hundred or so Google searches*: In 2009, an official Google blog estimated that one search used about one kilojoule of energy: https://googleblog.blogspot.com/2009/01/powering-google-search.html. To make a cup of tea means raising the temperature of 6.75 ounces of water by 135°F, which requires about 70 KJ. I've assumed Google's technology has become more efficient since 2009 and hence I say "a hundred or so Google searches."

167　*According to Google's own data*: See "Google Environmental Report 2019."

167　*small country like Lithuania*: See https://www.cia.gov/library/publications/the-world-factbook/fields/253rank.html.

167　*use about 1 percent of global electricity*: See "How to Stop Data Centres from Gobbling up the World's Electricity," *Nature*, September 12, 2018.

167　*2 percent of the world's carbon emissions*: See above.

167　*20 percent of the world's electricity*: See "On Global Electricity Usage of Communication Technology: Trends to 2030" by Anders S. G. Andrae and Tomas Edler, *Challenges* 6 (1) (2015): 117–57.

167　*The cooling systems*: See "Computer Engineering: Feeling the Heat," *Nature* 492 (December 13, 2012).

168　*Ma Bell, as the company*: For a detailed history of Bell Labs, see *The Idea Factory: Bell Labs and the Great Age of American Innovation* by Jon Gertner.

169　*"Mr. Watson, come here"*: See above.

170　*Claude Shannon was born*: For details on Shannon's life and work, see *A Mind at Play: How Claude Shannon Invented the Information Age* by Jimmy Soni and Rob Goodman.

170　*"if you walked a couple of blocks"*: As quoted in "Profile of Claude Shannon" by Anthony Liversidge, in the introduction to *Claude Elwood Shannon: Collected Papers*.

170　*"when a husband was capable"*: As quoted in *Mind at Play* by Soni and Goodman.

171　*"A decidedly unconventional type of youngster"*: Letter from Vannevar Bush to E. B. Wilson, December 15, 1938.

171　*"Bix Beiderbecke, you got him?"*: As quoted in *Mind at Play* by Soni and Goodman.

172　*"Christlike looks"*: As quoted in *Idea Factory* by Gertner.

172　*"How can you be anything else?"*: As quoted in *Mind at Play* by Soni and Goodman.

172　*Shannon loved poetry*: Interview with Norma Levor in the film, *The Bit Player*.

172　*"So sweet, so full of fun"*: As quoted in *Mind at Play* by Soni and Goodman.

172　*"scaring me shitless"*: As quoted in *Idea Factory* by Gertner.

172　*"I tried to get him to go to an analyst"*: See above.

173　*"there is not much"*: From "Sigsaly Story" by Patrick D. Weadon, from the NSA website.

174　*"Turing and I had an awful lot in common"*: From "Claude E. Shannon: An Interview Conducted by Robert Price," July 28, 1982.

174　*"We had dreams"*: From an interview Shannon gave to Friedrich-Wilhelm Hagemeyer, February 28, 1977.

174　*"I had talked to him several times"*: From "Shannon: An Interview by Price."

174　*"They never told me"*: As quoted in *Mind at Play* by Soni and Goodman.

174　*"A Mathematical Theory of Communication"*: From *Bell System Technical Journal* 27 (1948).

175　*"reproducing at one point"*: From the above paper.

180　*Shannon pointed the similarity out to John von Neumann*: This anecdote originates in a 1971 article, "Energy and Information," *Scientific American*, by Myron Tribus and

Edward C. McIrvine. But in 1982, in a taped interview Shannon is rather hazy about why he chose the term *entropy*.

181  *MST PPL HV*: From "Information Theory" by Claude E. Shannon, *Encyclopaedia Britannica*, 14th ed.

183  *"information is physical"*: Landauer wrote an article with this title in 1991 in *Physics Today*.

183  *another Bell Labs discovery*: See *Idea Factory* by Gertner.

184  *10 million-millionths of a joule of heat*: See the calculation in chapter 10 of *The Logician and the Engineer: How George Boole and Claude Shannon Created the Information Age* by Paul J. Nahin.

184  *hot plates on stoves*: See "A Research Agenda Towards Zero-Power ICT" by Gabriel Abadal Berini, Giorgos Fagas, Luca Gammaitoni, and Douglas Paul, 2014.

184  *exhaust nozzle of a rocket*: See above.

185  *1.43 billion tons of carbon*: SMART 2020: Enabling the Low Carbon Economy in the Information Age, a report by the Climate Group on behalf of the Global eSustainability Initiative (GeSI).

## Chapter Seventeen: Demons

187  *The process of diffusion*: From *The Kinetic Theory of the Dissipation of Energy* by William Thomson, 1874.

187  *planned to write a history*: The book would be called *Sketch of Thermodynamics* and was first published in 1868.

187  *"Clausius and others have cut up very rough"*: Letter from Tait to Maxwell, December 6, 1867.

187  *"I could make no assertions"*: Letter from Maxwell to Tait, December 11, 1867.

188  *came up with a thought experiment*: See above.

188  *"Conceive a finite being"*: See above.

189  *"No work has been done"*: See above.

189  *"We can't, not being clever enough"*: See above.

189  *William Thomson, who in an 1874 paper*: "The Kinetic Theory of the Dissipation of Energy," *Proceedings of the Royal Society of Edinburgh* 8 (1874).

190  *He had also, in his PhD thesis*: "On the Extension of Phenomenological Thermodynamics to Fluctuation Phenomena." In 1925 it was published in *Zeitschrift für Physik* 32.

190  *"The Decrease of Entropy by Intelligent Beings"*: Published in 1929 in *Zeitschrift für Physik* 53.

192  *"a perpetual-motion machine"*: See above.

192  *he was vague about how his demon causes entropy to increase*: See section 1.3 in *Maxwell's Demon 2: Entropy, Classical and Quantum Information, Computing* edited by Harvey Leff and Andrew F. Rex; and "Demons, Engines and the Second Law" by Charles H. Bennett, *Scientific American*, November 1987.

192  *The few scientific papers*: Examples include a 1951 paper by Leon Brillouin, "Maxwell's Demon Cannot Operate: Information and Entropy," and a 1964 paper by Dennis Gabor, "Light and Information."

192  *"We are looking for general laws"*: From "The Fundamental Physical Limits of Computation" by Charles H. Bennett and Rolf Landauer, *Scientific American*, July 1985.

193  *The senior figure of the two, Rolf Landauer*: See *Rolf W. Landauer, 1927–1999: A Biographical Memoir* by Charles H. Bennett and Alan B. Fowler.

193   *An early machine*: See "Ballistic Research Laboratories Report no. 971," December
      1955, US Department of Commerce, Office of Technical Services; and *The History of
      the ENIAC Computer* by Mary Bellis, ThoughtCo, February 11, 2020.
193   *IBM brought out its first transistor-based computer*: This was the IBM 7070.
193   *Power consumption for*: "IBM 7070 Data Processing System" by J. Svigals, *Proceedings
      of the Western Joint Computer Conference*, 1959.
194   *"The search for faster"*: "Irreversibility and Heat Generation in the Computing Pro-
      cess" by R. Landauer, *IBM Journal*, July 1961.
194   *Then in 1972, twenty-nine-year-old Charles Bennett*: Biographical information on
      Bennett is from the IBM website.
194   *To see how, picture again Leo Szilard's demon*: See "Demons, Engines" by Bennett.
196   *3,000 billion-billionths of a joule*: See section 10.4 of *Logician and the Engineer* by Nahin.
196   *Eric Lutz and his colleagues*: See "Experimental Verification of Landauer's Principle
      Linking Information and Thermodynamics" by Antoine Bérut, Artak Arakelyan,
      Artyom Petrosyan, Sergio Ciliberto, Raoul Dillenschneider, and Eric Lutz, *Nature*
      483 (March 2012).
196   *albeit with one caveat*: For a sense of the immense difficulties involved in building a
      computer that didn't dissipate energy, see chapter 7 from "Introduction to Nanoelec-
      tronics," MIT OpenCourseWare.
197   *By measuring how rapidly the cells reproduce*: "Minimum Energy of Computing, Fun-
      damental Considerations" by Victor Zhirnov, Ralph Cavin, and Luca Gammaitoni,
      chapter from the book *ICT—Energy—Concepts Towards Zero—Power Information
      and Communication Technology*.

## Chapter Eighteen: The Mathematics of Life

199   *a mathematical model*: From "The Chemical Basis of Morphogenesis" by Alan Tur-
      ing, *Philosophical Transactions of the Royal Society of London*, Series B 237 (1952–54).
199   *He is best known for his pivotal role*: See a range of biographies including *The Man
      Who Knew Too Much: Alan Turing and the Invention of the Computer* by David
      Leavitt and *Alan Turing: The Enigma* by Andrew Hodges.
200   *"There should be no question in"*: See *Cryptographic History of Work on the German
      Naval Enigma* by Hugh Alexander.
200   *rightly celebrated*: For example, *Breaking the Code*, a play by Hugh Whitmore; *Brit-
      ain's Greatest Codebreaker*, broadcast on UK Channel 4; and *The Imitation Game*,
      film starring Benedict Cumberbatch.
200   *Alan Mathison Turing was born*: As well as the biographies mentioned above, see
      *Alan M. Turing* by his mother, Sara Turing; and *Alan Turing: The Life of a Genius* by
      his nephew Dermott Turing.
200   *"accepted procedure for those"*: See "My Brother Alan" by Turing's brother, John,
      which appears at the end of their mother's book.
200   *"clever boys and hardworking boys, but Alan is a genius"*: From *Alan M. Turing* by Turing.
200   About a Microscope: See above.
200   Natural Wonders Every Child Should Know: A book by Edwin Tenney Brewster.
201   *"How they find out when"*: From above.
201   *"Hockey, or Watching the Daisies Grow"*: Sara Turing drew the picture and sent it to
      the matron at Turing's school in 1923.

201 *"On Computable Numbers"*: "On Computable Numbers, with an Application to the Entscheidungsproblem" by Alan Turing, first published in *Proceedings of the London Mathematical Society*, ser. 2, 42 (1936–37).

201 *Most historians now regard Turing's Universal Machine*: See chapter 6 of *The Turing Guide* by B. Jack Copeland.

202 *a fifteen-year-old Jewish refugee named Robert Augenfeld*: See *Alan Turing* by Hodges and a brief essay by Augenfeld that was written shortly before he died.

202 *"vapid conversation"*: See "My Brother Alan" by Turing's brother.

203 *he told her of his "homosexual tendencies"*: See above.

203 *The head of the daisy*: See "The Mathematical Daisy" by Robert Dixon, *New Scientist*, December 17, 1981.

204 *Turing played a crucial role*: See chapter 6 of *Turing Guide* by Copeland; chapter 9 of *The Essential Turing* by B. Jack Copeland; and Manchester University website http://cura tion.cs.manchester.ac.uk/computer50/www.computer50.org/mark1/new.baby.html.

204 *"Computing Machinery and Intelligence"*: The article first appeared in *Mind* 59 (1950).

205 *"The Chemical Basis of Morphogenesis"*: The article first appeared in *Philosophical Transactions of the Royal Society of London, Series B* 237 (1952–54).

205 *Turing considered it his best work*: See an unpublished short story by Turing at the Turing Digital Archive, maintained by King's College Cambridge, ref. AMT/A 13.

206 *"a possible mechanism by which the"*: "Chemical Basis of Morphogenesis," *Philosophi cal Transactions*.

206 *"Just think of a drop of ink in water"*: See "Positional Information and Reaction-Diffusion: Two Big Ideas in Developmental Biology Combine" by Jeremy B. A. Green and James Sharpe, *Development* 142 (2015): 1203–11.

207 *"Such a system"*: "Chemical Basis of Morphogenesis," *Philosophical Transactions*.

207 *"This model will be a simplification"*: See above.

208 *In unpublished notes*: See Turing Digital Archive, AMT/C/27 image 014.

209 *"that one cannot hope to"*: "Chemical Basis of Morphogenesis," *Philosophical Transactions*.

210 *"a continual supply of free energy"*: See above.

211 *Turing wrote the beginnings of a short story*: Sometimes called "Pryce's Buoy," at Turing Digital Archive, AMT/A 13.

212 *"I've got a shocking tendency"*: Letter from Turing to Gandy, March 11, 1953, Turing Digital Archive, AMT/D/4.

212 *Experts on the male reproductive system*: See interview with Professor Allan Pacey, University of Sheffield, in *Britain's Greatest Codebreaker*.

212 *"It will be difficult, in some places"*: Turing Digital Archive, AMT/C 27.

212 *"a deliberate act"*: The coroner was quoted in the *Daily Telegraph*, June 11, 1954.

213 *"If I had so much as parked my bicycle"*: Letter from Turing to Norman Routledge, Turing Digital Archive, AMT/D 14a.

213 *According to John's son, Dermot Turing*: From interview with Dermot Turing in *Britain's Greatest Codebreaker*.

214 *Turing's concept of spontaneous pattern formation*: For an excellent article discussing this and alternative theories, see "Positional Information and Reaction-Diffusion" by Green and Sharpe.

215 *"the antithesis of positional information"*: See "Positional Information and Pattern Formation" by Lewis Wolpert, *Current Topics in Developmental Biology* 6 (1971).

215  *They found compelling evidence*: See "Positional Information Revisited" by Lewis Wolpert, *Development*, 1989.

216  *By studying mice, a team of researchers in Germany*: See "WNT and DKK Determine Hair Follicle Spacing through a Reaction-Diffusion Mechanism" by Stefanie Sick, Stefan Reinker, Jens Timmer, and Thomas Schlake, *Science* 01 (December 2006).

216  *One, by Professor Jeremy Green and his team of developmental biologists*: "Periodic Stripe Formation by a Turing-Mechanism Operating at Growth Zones in the Mammalian Palate" by Andrew D. Economou, Atsushi Ohazama, Thantrira Porntaveetus, Paul T. Sharpe, Shigeru Kondo, M. Albert Basson, Amel Gritli-Linde, Martyn T. Cobourne, and Jeremy B. A. Green, *Nature Genetics*, February 19, 2012.

217  *This rugae paper was followed two years*: "Digit Patterning Is Controlled by a Bmp-Sox9-Wnt Turing Network Modulated by Morphogen Gradients" by J. Raspopovic, L. Marcon, L. Russo, and J. Sharpe, *Science*, February 2014.

217  *described Turing as a "genius"*: See interview on YouTube, "Lewis Wolpert—Reaction Diffusion Theory That Goes Back to Alan Turing," October 2017.

## Chapter Nineteen: Event Horizon

219  *Bekenstein and Hawking were the first*: From *The Black Hole War: My Battle with Stephen Hawking to Make the World Safer for Quantum Mechanics* by Leonard Susskind.

219  *Your idea is so crazy*: *Geons, Black Holes, and Quantum Foam: A Life in Physics* by John Archibald Wheeler with Kenneth Ford.

220  *"happiest thought of my life"*: From *The Collected Papers of Albert Einstein, Vol. 7: The Berlin Years*.

221  *"For an observer in free fall"*: From *Collected Papers of Albert Einstein, Vol. 7*.

222  *"That gravity should be"*: From a letter from Isaac Newton to Richard Bentley, 1692–93.

225  *In early 1916*: "Über das Gravitationsfeld eines Massenpunktes nach der Einsteinschen Theorie" by Karl Schwarzschild, February 1916.

226  *evocative nickname*: black holes: During a 1967 talk at NASA's Goddard Institute, the lecturer John Wheeler asked the audience to suggest a shorter term to describe a "gravitationally completely collapsed object." Someone offered *black hole*. From *Geons, Black Holes* by Wheeler with Ford.

226  *Imagine a shallow ocean of water*: I have taken this analogy from Leonard Susskind's *Black Hole War*. In this book, Susskind credits physicist Bill Unruh with originating it.

228  *That's why astronomers have observed stars*: "Investigating the Relativistic Motion of the Stars Near the Supermassive Black Hole in the Galactic Center" by M. Parsa1, A. Eckart, B. Shahzamanian, V. Karas, M. Zajaček, J. A. Zensus, and C. Straubmeier, *Astrophysical Journal*, 2017.

229  *These waves were found*: See "Einstein's Gravitational Waves Found at Last," *Nature*, February 2016.

229  *"They were intelligent enough"*: Hawking's physics tutor Robert Berman is quoted as saying this in the *New York Times Magazine*, January 23, 1983.

230  *Hawking published a paper*: This is known as Hawking's area theorem.

230  *a black hole cannot radiate heat and therefore cannot have entropy*: See *A Brief History of Time: From the Big Bang to Black Holes* by Stephen Hawking.

230  *Jacob Bekenstein and his supervisor, John Wheeler*: For details of this meeting and more biographical information on both, see *Of Gravity, Black Holes and Information*

by Jacob D. Bekenstein and *Geons, Black Holes* by Wheeler with Ford. Also see *Black Holes and Time Warps: Einstein's Outrageous Legacy* by Kip S. Thorne.

231  *popularized the term* black hole: See note above.

231  *"A few actually flew"*: From *Of Gravity, Black Holes* by Bekenstein.

231  *"As a young man"*: See above.

232  *"I always feel like a criminal"*: From *Geons, Black Holes* by Wheeler with Ford.

232  *"I was very dissatisfied"*: From *Of Gravity, Black Holes* by Bekenstein.

233  *"Bekenstein's style of doing physics"*: From *Black Hole War* by Susskind.

234  *generalized second law of thermodynamics*: See "Black Holes and Entropy" by Jacob D. Bekenstein, *Physical Review D* 7 (8) (April 15, 1973).

234  *"Your idea is so crazy"*: See second note of this chapter.

234  *"Those were two lonely years"*: *Of Gravity, Black Holes* by Bekenstein.

234  *"I must admit that"*: *Brief History of Time* by Hawking.

235  *"surprise and annoyance"* and *"afraid that if Bekenstein"*: See above.

235  *Hawking's exact reasoning is tricky*: A complete understanding would require a hitherto undiscovered theory combining general relativity and quantum mechanics. Hawking's work is a crucial step toward that theory.

236  *"It turned out in the end"*: See above.

237  *That field is information theory*: For an excellent discussion of this, see "Information in the Holographic Universe" by Jacob D. Bekenstein, *Scientific American*, August 2003.

238  *The reason is that in 1998*: See "Observational Evidence from Supernovae for an Accelerating Universe and a Cosmological Constant," *Astronomical Journal*, September 1998; and "The Acceleration of the Expansion of the Universe: A Brief Early History of the Supernova Cosmology Project" by Gerson Goldhaber, *AIP Conference Proceedings*, 2009.

239  *so, too, is gravity*: See "The Illusion of Gravity" by Juan Maldecena, *Scientific American*, April 1, 2007.

240  *"We are just an advanced breed of monkeys"*: From an interview Hawking gave to *Der Spiegel* in 1988.

## Epilogue

242  *Victorian scientist—John Tyndall*: See *John Tyndall: Essays on a Natural Philosopher* edited by W. H. Brock, N. D. McMillan, and R. C. Mollan; and "On the Origin of 'the Greenhouse Effect': John Tyndall's 1859 Interrogation of Nature" by Mike Hulme, Royal Meteorological Society, 2009.

242  *Tyndall devised a beautiful and historic experiment*: See "On the Absorption and Radiation of Heat by Gases and Vapours, and on the Physical Connexion of Radiation, Absorption and Conduction" by John Tyndall, Bakerian Lecture, 1861.

244  *why, as early as 1917, Alexander Graham Bell*: From "Some of the Problems Awaiting Solution," an address by Alexander Graham Bell at the McKinley Manual Training School, Washington, D.C., February 1, 1917.

244  *the United Kingdom's electricity*: UK Department for Business, Energy & Industrial Strategy (2020).

244  *James Lovelock*: See article by Lovelock in the *Independent*, May 24, 2004.

244  *Mark Lynas*: See *Nuclear 2.0: Why a Green Future Needs Nuclear Power* by Mark Lynas.

# Bibliography

The book is, roughly speaking, in three parts. Below are the texts I found invaluable for each section.

Part One:
The Discovery of Energy and Entropy—
Chapters One to Four

*Against Intellectual Monopoly* by Michele Boldrin and David K. Levine
*The Analytical Theory of Heat* by Joseph Fourier
*De l'Angleterre et des Anglais* by Jean-Baptiste Say
*Degrees Kelvin* by David Lindley
*The Edge of Objectivity: An Essay in the History of Scientific Ideas* by Charles Coulston Gillespie
*Energy and Empire: A Biographical Study of Lord Kelvin* by Crosbie Smith and M. Norton Wise
*Energy, the Subtle Concept* by Jennifer Coopersmith
*From Watt to Clausius* by D. S. L. Cardwell
*Great Physicists* by William H. Cropper
*Inventing Temperature: Measurement and Scientific Progress* by Hasok Chang
*James Joule: A Biography* by D. S. L. Cardwell
*James Prescott Joule* by Osborne Reynolds
*Jean-Baptiste Say: Revolutionary, Entrepreneur, Economist* by Evert Schoorl
*Lord Kelvin: An Account of His Scientific Life and Work* by Andrew Gray
*The Lunar Men: The Friends Who Made the Future* by Jenny Uglow
*Mathematical and Physical Papers*, vols. 1–3, by Sir William Thomson
*Modern Engineering Thermodynamics* by Robert T. Balmer
*The Oxford Handbook of the History of Physics* by Jed Buchwald and Robert Fox
*Popular Lectures and Addresses* by Sir William Thomson
*Reflections on the Motive Power of Fire* by Sadi Carnot, translated and edited by Robert Fox. This contains a highly informative introduction by Fox.
*Reflections on the Motive Power of Heat* by Sadi Carnot, edited by R. H. Thurston. This version contains a memoir of Sadi by his brother, Hippolyte, and extracts from Sadi's unpublished writings.
*The Science of Energy* by Crosbie Smith

283

*Scientific Papers* by James Joule
*Song of the Clyde* by Fred M. Walker
*Theory and Construction of a Rational Heat Motor* by Rudolf Diesel
*The Unbound Prometheus: Technological Change and Industrial Development in Western Europe from 1750 to the Present* by David S. Landes
*When Physics Became King* by Iwan Rhys Morus

Part Two:
Classical Thermodynamics—
Chapters Five to Twelve

*Aesthetics, Industry, and Science: Hermann von Helmholtz and the Berlin Physical Society* by M. Norton Wise
*Black-Body Theory and the Quantum Discontinuity, 1894–1912* by Thomas S. Kuhn
*Boltzmann's Atom: The Great Debate That Launched a Revolution in Physics* by David Lindley
*The Economic Development of France and Germany, 1815–1914* by J. H. Clapham
*From Eternity to Here: The Quest for the Ultimate Theory of Time* by Sean Carroll
*The German Genius* by Peter Watson
*Helmholtz: A Life in Science* by David Cahan
*Hermann Ludwig Ferdinand von Helmholtz* by John Gray McKendrick
*Hermann von Helmholtz* by Leo Koenigsberger, translated by Frances A. Welby
*Intellectual Mastery of Nature: Theoretical Physics from Ohm to Einstein*, vols. 1 and 2, by Christa Jungnickel and Russell McCormmach
*Josiah Willard Gibbs: The History of a Great Mind* by Lynde Phelps Wheeler
*Kinetic Theory*, vols. 1 and 2, by Stephen G. Brush
*Lectures on Gas Theory* by Ludwig Boltzmann, translated by Stephen G. Brush
*The Life of James Clerk Maxwell* by Lewis Campbell
*Life's Ratchet: How Molecular Machines Extract Order from Chaos* by Peter M. Hoffmann
*Lord Kelvin and the Age of the Earth* by Joe D. Burchfield
*Ludwig Boltzmann: The Man Who Trusted Atoms* by Carlo Cercignani
*The Man Who Changed Everything: The Life of James Clerk Maxwell* by Basil Mahon
*The Mechanical Theory of Heat, with Its Applications to the Steam Engine and to the Physical Properties of Bodies* by Rudolf Clausius
*On the Origin of Species* by Charles Darwin
*Populäre Schriften* by Ludwig Boltzmann
*Refrigeration: A History* by Carroll Gantz
*The Scientific Papers of J. Willard Gibbs*, vols. 1 and 2
*The Second Physicist: On the History of Theoretical Physics in Germany* by Christa Jungnickel and Russell McCormmach
*Willard Gibbs* by Muriel Rukeyser

Part Three:
The Consequences of Thermodynamics—
Chapters Thirteen to Nineteen

*Alan Turing: The Enigma* by Andrew Hodges
*Alan Turing: The Enigma Man* by Nigel Cawthorne

*Alan Turing: The Life of a Genius* by Dermot Turing

*The Black Hole War: My Battle with Stephen Hawking to Make the World Safer for Quantum Mechanics* by Leonard Susskind

*Black Holes and Time Warps: Einstein's Outrageous Legacy* by Kip S. Thorne

*A Brief History of Time: From the Big Bang to Black Holes* by Stephen Hawking

*The Bumpy Road: Max Planck from Radiation Theory to the Quantum, 1896–1906* by Massimiliano Badino

*Einstein: His Life and Universe* by Walter Isaacson

*Einstein and the Quantum: The Quest of the Valiant Swabian* by A. Douglas Stone

*The Einstein Theory of Relativity: A Concise Statement* by H. A. Lorentz

*Einstein's Berlin: In the Footsteps of a Genius* by Dieter Hoffmann

*Einstein's Masterwork: 1915 and the General Theory of Relativity* by John Gribbin

*Emmy Noether, 1882–1935* by Auguste Dick, translated by H. I. Blocher

*Emmy Noether's Wonderful Theorem* by Dwight E. Neuenschwander

*The Essential Turing: Seminal Writings in Computing, Logic, Philosophy, Artificial Intelligence and Artificial Life: Plus the Secrets of Enigma* edited by B. Jack Copeland

*Genius in the Shadows* by William Lanouette with Bela Silard

*Geons, Black Holes, and Quantum Foam: A Life in Physics* by John Archibald Wheeler with Kenneth Ford

*Of Gravity, Black Holes and Information* by Jacob D. Bekenstein

*The Idea Factory: Bell Labs and the Great Age of American Innovation* by Jon Gertner

*Information Theory: A Tutorial Introduction* by James V. Stone

*Information Theory and Evolution* by John Scales Avery

*An Institute for an Empire: The Physikalisch-Technische Reichsanstalt, 1871–1918* by David Cahan

*An Introduction to Black Holes, Information and the String Theory Revolution: The Holographic Universe* by Leonard Susskind and James Lindesay

*An Introduction to Information Theory: Symbols, Signals and Noise* by John R. Pierce

*The Innovators* by Walter Isaacson

*Life and Scientific Work of Peter Guthrie Tate* by Cargill Gilston Knott

*The Logician and the Engineer: How George Boole and Claude Shannon Created the Information Age* by Paul J. Nahin

*Lonely Hearts of the Cosmos: The Story of the Scientific Quest for the Secret of the Universe* by Dennis Overbye

*The Man Who Knew Too Much: Alan Turing and the Invention of the Computer* by David Leavitt

*Maxwell's Demon 2: Entropy, Classical and Quantum Information, Computing* edited by Harvey Leff and Andrew F. Rex

*A Mind at Play: How Claude Shannon Invented the Information Age* by Jimmy Soni and Rob Goodman

*Planck's Original Papers in Quantum Physics* translated by D. ter Haar and Stephen G. Brush

*Quantum: Einstein, Bohr and the Great Debate about the Nature of Reality* by Manjit Kumar

*Quantum Profiles* by Jeremy Bernstein

*17 Equations That Changed the World* by Ian Stewart

*Significant Figures: Lives and Works of Trailblazing Mathematicians* by Ian Stewart

*Sketch of Thermodynamics* by Peter Guthrie Tate

*Stephen Hawking: His Life and Work* by Kitty Ferguson

*Stephen Hawking's Universe: An Introduction to the Most Remarkable Scientist of Our Time* by John Boslough

*A Student's Guide to Einstein's Major Papers* by Robert E. Kennedy

*Symmetry and the Beautiful Universe* by Leon M. Lederman and Christopher T. Hill

*The Theory of Relativity and Other Essays* by Albert Einstein

*Three Degrees above Zero* by Jeremy Bernstein

*The Turing Guide* by B. Jack Copeland, Jonathan P. Bowen, Mark Sprevak, Robin Wilson, and others

## Epilogue

*John Tyndall: Essays on a Natural Philosopher* edited by W. H. Brock, N. D. McMillan, and R. C. Mollan

# Index

# About the Author

**Paul Sen** first encountered thermodynamics while studying engineering at the University of Cambridge. He became a documentary filmmaker with a passion for communicating profound scientific ideas in an engaging and accessible way to a wide audience with landmark TV series such as *Triumph of the Nerds* and *Atom*, which brought a love of storytelling to the worlds of science and technology. Paul's award winning TV company, Furnace, where he is creative director, has made BBC science series such as *Everything and Nothing*, *Order and Disorder*, and *The Secrets of Quantum Physics*, and ninety-minute films such as *Gravity and Me: The Force That Shapes Our Lives* and *Oak Tree: Nature's Greatest Survivor*. This won the prestigious Royal Television Society Award for best science and natural history program and the Grierson Award for best science documentary in 2016. This book is born out of Paul's conviction that the history of science is the history that matters.